SpringerBriefs in Applied Sciences and Technology

W0080197

For further volumes:
http://www.springer.com/series/8884

Jedol Dayou · Jackson Hian Wui Chang
Justin Sentian

Ground-Based Aerosol Optical Depth Measurement Using Sunphotometers

 Springer

Jedol Dayou
Jackson Hian Wui Chang
Justin Sentian
School of Science and Technology
Universiti Malaysia Sabah
Kota Kinabalu
Sabah
Malaysia

ISSN 2191-530X   ISSN 2191-5318 (electronic)
ISBN 978-981-287-100-8   ISBN 978-981-287-101-5 (eBook)
DOI 10.1007/978-981-287-101-5
Springer Singapore Heidelberg New York Dordrecht London

Library of Congress Control Number: 2014940151

Printed on acid-free paper

Springer is part of Springer Science+Business Media (www.springer.com)

Contents

1 Introduction 1
   1.1 Aerosol Basic 1
   1.2 Aerosol Impacts on Climate and Human Health 4
   1.3 Measurement of Aerosol Optical Depth 5
   1.4 AOD Measurement Using Sunphotometers 6
   References 7

2 Ground-Based Aerosol Optical Depth Measurements 9
   2.1 Theory of Aerosol Absorption and Scattering 9
      2.1.1 The Optical Properties of Spherical Particle 12
   2.2 Ground-Based Aerosol Optical Depth Retrieval 14
      2.2.1 Retrieval with Ground-Based Sunphotometry
             Radiometer 14
   2.3 Conventional Langley Calibration Method 18
   2.4 Historical Development of Langley Calibration Method 21
      2.4.1 Basic Sunphotometry Langley Method 21
      2.4.2 Circumsolar Langley Method 22
      2.4.3 Cloud-Screened Langley Method 23
      2.4.4 Maximum Value Composite (MVC) Langley Method ... 26
      2.4.5 Comparative Langley Method 28
   References 29

3 Near-Sea-Level Langley Calibration Algorithm 31
   3.1 Development of Near-Sea-Level Langley
       Calibration Algorithm 31
      3.1.1 Clear-Sky Detection Model 32
      3.1.2 Statistical Filter 34
   3.2 Implementation of the Proposed Calibration Algorithm 35
   References 36

4   Implementation of Perez-Dumortier Calibration Algorithm   39
   4.1   Instrumentation   39
   4.2   Determination of Langley Extraterrestrial Constant Using
        the Proposed Calibration Algorithm   41
   4.3   Retrieval of Spectral AOD   44
   4.4   Validation of the Proposed Calibration Algorithm   45
        4.4.1   Irradiance-Matched by i-SMARTS Radiative
                Transfer Code   45
        4.4.2   Radiative Closure Experiment   49
        4.4.3   Performance Analysis   49
   References   56

5   Conclusion   57
   5.1   Overview   57
   References   60

Index   61

Chapter 1
Introduction

Abstract The study of aerosol in atmospheric science has become an essential issue due to its ever-rising effects both in climate and human health. Continuous measurement of aerosol is hence necessary in order for better grasp of the effects. One of the practical methods to measure and monitor the aerosol content in atmosphere is by using sunphotometers. To embark on the issues surrounding the measurement using this device, some general knowledge regarding aerosol is first discussed in this chapter including some definitions related to aerosol, the importance of its measurement and monitoring, and the challenge faces the measurement using the device.

Keywords Airborne radiometer · Convective cloud · Manmade aerosol · Primary and secondary aerosols · Solar terrestrial radiation

1.1 Aerosol Basic

Aerosols are small solid particles or liquid droplets suspended in air or other gases environment. They can be naturally produced or manmade generated. Natural aerosols are emitted into the atmosphere by natural processes such as sea spray, volcanoes eruptions, windblown dust from arid and semi-arid regions, terrestrial biomass burning and others. Meanwhile, manmade aerosol are generated from combustion or emission from industrial, welding, and vehicle exhaust or produced intentionally for commercial uses (i.e. flame reactor aerosol that produces nano-particles). They have very limited life time of about a few days to one week. Despite their relatively short life times, they regularly travel over long distances via air trajectories. The transport pathways may vary seasonally and interannually depending on the air-mass altitude (Paul et al. 2011).

Aerosols have irregular shapes (i.e. aggregated, spherical, fibrous, and others), categorizing them is often based on the diameter of an idealized sphere, or better known as particle size. These sizes range from few nanometers to several tens of micrometers. More specifically, the aerosol particles with diameters $d \leq 0.1$ μm

J. Dayou et al., *Ground-Based Aerosol Optical Depth Measurement Using Sunphotometers*, SpringerBriefs in Applied Sciences and Technology, DOI: 10.1007/978-981-287-101-5_1, © The Author(s) 2014

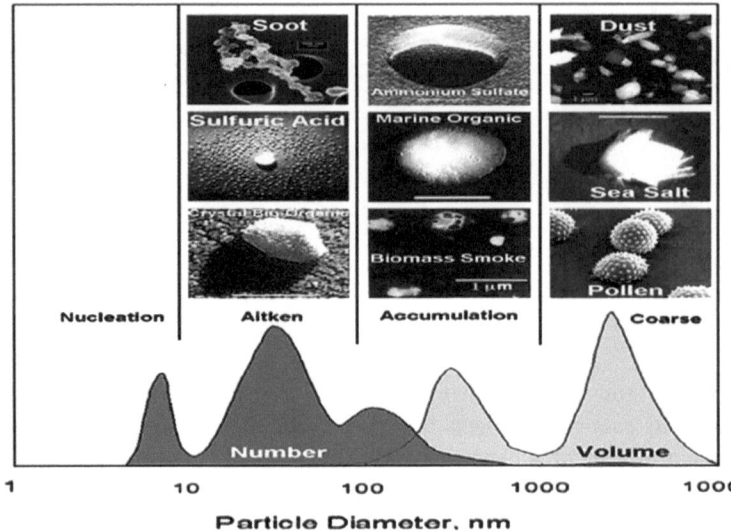

Fig. 1.1 Idealized number and volume distribution of atmospheric aerosols (Huang 2009)

belong to the nuclei mode, particles with diameter $0.1 \leq d \leq 2.5$ μm belong to the accumulation mode where all of these aerosol also known as fine particles, and particles with $d \geq 2.5$ μm are in the coarse mode. Aerosol particle of same size is known as monodisperse aerosol and this type of aerosol are normally produced in laboratory for specific purposes. Most aerosols particularly atmospheric aerosols are polydisperse, which have a range of particle sizes. Categorization of these aerosols is based on the use of the particle-size distribution.

Figure 1.1 shows the idealized number and volume density distribution of some atmospheric aerosols. The intermediate aerosol between nucleation and accumulation is Aietken mode, which makes up the majority of the aerosol mass.

Particles in this size range dominate aerosol direct interaction with sunlight of either scattering or absorbing. Particles at the small end of this size range play significant role in interactions with cloud, whereas particles at the large end contribute significantly near dust and volcanic sources, though of much less numerous. The particles of coarse mode are typically of very minor in number mass but high in volume distribution due to large particle size.

Aerosols may further be divided into two broad categories based on their nature of formation: primary and secondary aerosols. Primary aerosols are directly emitted as particles or liquid into the atmosphere by processes occurring on land or water which could be natural or manmade origin. Sources of primary aerosols are sea spray, windblown desert dust, volcanoes, plant particles, biomass burning, incomplete combustion of fossil fuels and etc. Secondary aerosols, on the other hand, are produced indirectly via atmospheric physical or chemical conversion of gases to particles compounds by nucleation and condensation gases precursors.

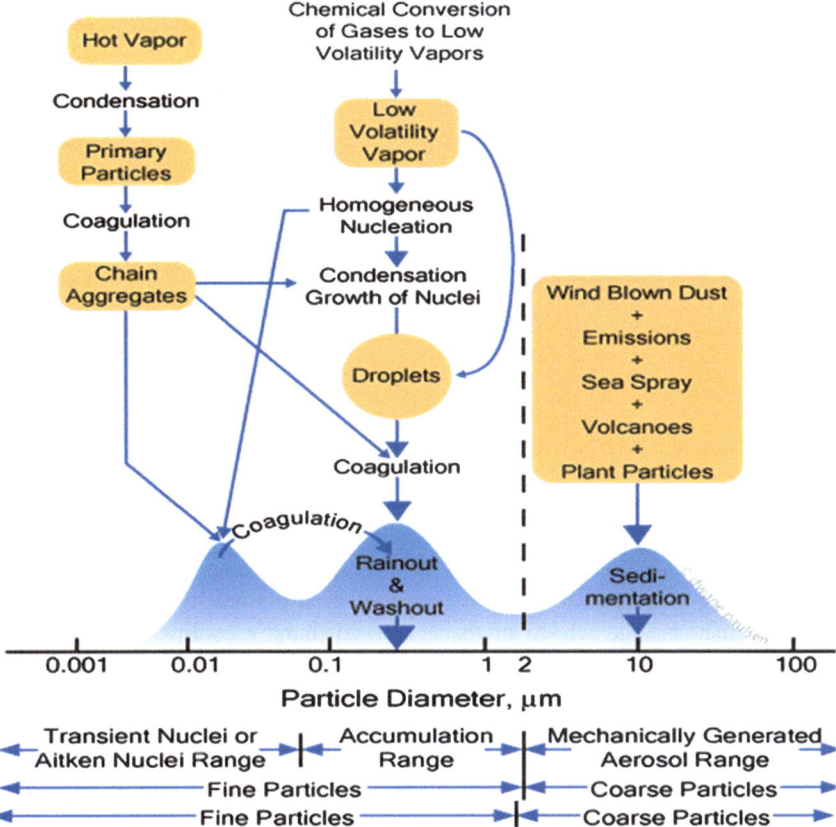

Fig. 1.2 Idealized schematic of the sources and sink of primary and secondary aerosols (Whitby et al. 1972)

They are mainly composed of sulphates, carbonaceous particles, nitrates, ammonium and mineral dust of industrial origin (Ghan and Schwartz 2007). Figure 1.2 depicts the atmospheric aerosol particle surface weighted by size distribution together with the different mechanisms of aerosol generation. The nuclei range is composed of both primary and secondary aerosols, but physical mechanisms such as condensation and coagulation quickly transform the particle mass from nuclei mode to accumulation mode. These mechanisms are related to their growth and may change their physical and chemical properties (Pöschl 2005). Besides, the sources and sink for the fine and coarse modes are also different. The fine particles are generally originated from the secondary aerosols and are deposited typically by rain-wash. Meanwhile, the coarse particles are mainly composed of primary aerosols and sink through sedimentation.

1.2 Aerosol Impacts on Climate and Human Health

Aerosols exert a variety of impacts on environment depending on their properties such as their concentration, size, structure, and chemical composition (Pöschl 2005). Unlike greenhouse gases, which possesses long life-time and a near-homogeneous spatial distribution, atmospheric aerosols are highly heterogeneous and have limited lifetime of the order of one week in the lower troposphere (Nair et al. 2012). This is because aerosols undergo various physical and chemical interactions and transformations in the atmosphere due to diffusion and aging processes such as nucleation, coagulation, humidification and gas to particle phase conversion (Chaâbane et al. 2005). These processes change their intrinsic characteristics and thus posing varying effects on environment. The two main concerns of aerosol effects are impacts on climate and human health, which are discussed in the following.

In general, aerosol effects on climate can be classified as direct and indirect with respect to radiative forcing of the climate system. Radiative forcing is changes in the energy flux of solar terrestrial radiation in the atmosphere, induced by anthropogenic or natural changes in atmospheric composition, earth surface properties, or solar activity. Firstly, most of aerosols are highly reflective and therefore increase the albedo of the earth and thereby cooling the surface and effectively offsetting greenhouse gas warming by about 25–50 % (Kiehl et al. 2000). This is described as the direct effect which makes the atmosphere brighter when viewed from space since much of Earth's surface is covered by dark oceans and aerosols also scatter visible light backing into space.

Secondly, aerosols in the low atmosphere can act as sites at which water vapor can accumulate during cloud droplet formation, serving as cloud condensation nuclei (CCN). Any changes in concentration or hydroscopic properties of such particle have potential to modify physical and radiative properties of cloud. In this case, the indirect effects of aerosol include an increase in cloud brightness, reduction in precipitation and increase in cloud lifetime. These indirect effects were first shown by Twomey (1974) that pollution can lead to an increase in solar radiation reflected by clouds. The influence of aerosol in this matter lies in the mechanism that the process of cloud condensation causes some of the particles in atmosphere to grow into cloud droplets. These growing particles have typically larger cross-sectional area than the nucleating particles. On the whole, the overall effect is a great magnification of the light scattering power of those particles and resulting in a negative radiative forcing at top of atmosphere (TOA) (Lohmann 2006).

The scattering and absorption of radiation by aerosols can also cause perturbation in Earth's energy balance in a semi-direct effect (Yu et al. 2006). The effects of this are twofold: warming the atmosphere and cooling the surface below. For instance, black carbon or biomass burning aerosols are absorbing aerosols that absorb incident sunlight and re-radiate at infrared wavelength to cause positive radiative forcing and contributing to global warming (Mishchenko et al. 2007). In contrast, negative radiative forcing type aerosols are sulphate, nitrate and organic carbon particles which causes atmospheric and surface cooling by reflecting solar radiation back to space (Myhre et al. 2009). In this way, an overall effect includes of reducing the atmosphere vertical temperature gradient and therefore contributing to the reduction of formation of convective cloud.

Aerosol are also highly interactive with other components of the climate system, for instance, acidification of lakes and forests through the deposition of sulfates and nitrates and reduction of snow and ice albedo through the deposition of black carbon (Ghan and Schwartz 2007). Also reported in renewable energy application is the most important variable that conditions the accuracy of the predicted spectra under cloudless skies is aerosol optical depth (AOD) (Gueymard 2008), which directly constitutes the performance of solar photovoltaic technology.

Excessive inhalation of particulate matter by human is detrimental to asthma, lung cancer, cardiovascular issues, birth defects, and more severely premature death. Large particles are typically filtered in the nose and throat via cilia or mucus but smaller $PM_{10 \ \mu m}$ can penetrate to the deepest part of lung and settle in there to cause the adverse effects. Long-term exposure to combustion-related fine particle air pollution is an important environmental factor for cardiopulmonary and lung cancer mortality (Arden Pope III et al. 2002). Each 10 $\mu g/m^3$ elevation in fine particle in air pollution was associated with approximately a 4 %, 6 % and 8 % increased risk of all-cause, cardiopulmonary and lung cancer mortality respectively. At shorter time exposure (<24 h), these particles may induce plaque rupture and activate blood platelets, leading to acute peripheral arterial events such as myocardial infarction (Emmerechts et al. 2011).

Ultrafine (UF) particles (<2.5 μm), the smallest pollutant particles are even more dangerous due to their high content of organic chemicals and prooxidative potentials. Test subjects exposed to UF particles exhibited significantly larger early atherosclerotic lesions and also resulted in an inhibition of anti-inflammatory capacity of plasma high-density lipoprotein (Araujo et al. 2008). This is due to the fact that UF particles concentrate the proatherogenic effect of ambient PM and constitute a significant cardiovascular risk factor.

1.3 Measurement of Aerosol Optical Depth

With regard to the effects of aerosol on climate and human heath, aerosol must be accurately monitored in mass concentration, particle size and size-dependent composition, optical properties, solubility and the ability to serve as nuclei of cloud particles (Ghan and Schwartz 2007). Accurate measurement of these parameters renders not only the comprehension of their effects on environment but also permits large-scale representing them in numerical models.

One of the most important parameters in aerosol measurement is aerosol optical depth (AOD). It is an optical parameter that represents the magnitude of depletion of solar insolation due to scattering and absorbing processes caused by aerosols. Besides, the spectral AODs also have an imprint of the aerosol columnar size distributions (CSDs) that provides a rough estimation of type of aerosols (Satheesh et al. 2005). Since decades ago, AOD monitoring had been introduced in many environmental studies. In general, there are four methods commonly used in AOD measurement which are retrieval with satellite data, ground-based sunphotometry radiometer, airborne radiometer and lidar.

Among them, satellite data is most frequently used due to its large spatial resolution. Strengths of satellite approach are also not limited to emission iden- tification, filling gaps in areas where no ground sensors, defining production, oxidation, and evolution process from biomass burning. However its accuracy is always under much debate due to improper treatment of the reflection and aerosol models used in the AOD inversion algorithm (Remer et al. 2005). The uncertainty is even more significant over areas where satellite overpass does not coincide with the area or period of interest. For airborne radiometer and lidar retrieval, they are able to derive multiple values of AOD vary with altitude and also most flexible in terms of measurement time. Yet they are not the priority in AOD monitoring because they require complex instrumentation and the high maintenance cost is always an important issue for areas where technical and financial support are acutely limited.

1.4 AOD Measurement Using Sunphotometers

In contrast, retrieval with ground-based sunphotometry radiometer has good accuracy, provides highest resolution in spectral and temporal, as well as cost- effective in financial wise when compared to the other methods. Therefore, it is often a preferable selection for application that requires high accuracy mainly for cross-validation purposes and areas where inspection of AOD is still at its beginning stage. However, problems associated to this method do exist particularly in the calibration issue.

Conventionally, calibration is performed using standard laboratory lamps. It is also known as absolute calibration where the determination of the absolute response of a spectrometer is for a given spectral irradiance incident on the instrument. These lamps typically have inconsistent uncertainty ranging from 1 to 4 % in the wavelength from 400 to 1070 nm (Kiedron et al. 1999). They are also prohibitive with necessary power supplies, fragile and have a limited lifetime of about 50 h (Slusser et al. 2000). In contrast, an alternative to absolute calibration known as Langley method is a passive calibration procedure that uses solar radiation as the light source. It is performed using solar disc irradiances to determine the instrument output at top-of-atmosphere and subsequently divide this output by spectrally extraterrestrial irradiances. Detail of the principle is discussed in Sect. 2.5.

Though Langley method is economical, it is always complicated by the possible temporal drifts in the atmospheric condition during the calibration period (Shaw 1983). In AERONET, one of the most established aerosol monitoring networks, the reference instruments are typically recalibrated on a basis of 2–3 months cycle in high altitude (3400 m) condition at Mauna Loa Observatory (MLO), for clear and aerosol-stable atmosphere. In addition to this, the possible degradation of the instrument itself may also hinder the performance of the retrieval (Nieke et al. 1999; Schmid and Wherli 1994). Thus, frequent calibration is necessary to ensure correct ground-truth measurements, especially for monitoring of long-term variation of atmospheric turbidity (Arai and Liang 2011). However, regular access to high altitude for periodic calibration is not efficient in terms of accessibility and economical prospects. Therefore, most instruments are calibrated against a reference instrument with a MLO-derived extrapolated value (Saeed and Al-Dashti 2010) but these secondary calibrated instruments typically have larger uncertainties than the reference instrument uncertainty (Holben et al. 1998). This has created the needs for calibration protocol that can be performed at any altitude point and instance.

References

Arai K, Liang XM (2011) Comparative calibration method between two different wavelengths with aureole observations at relatively long wavelength. Int J Appl Sci 2:93–101

Araujo Ja, Barajas B, Kleinman M et al (2008) Ambient particulate pollutants in the ultrafine range promote early atherosclerosis and systemic oxidative stress. Circ Res 102:589–96

Arden Pope III C, Burnett RT, Thun MJ et al (2002) Lung cancer cardiopulmonary mortality, and long-term exposure to fine particulate air pollution. J Am Med Assoc 287:1132–1141

Chaâbane M, Masmoudi M, Medhioub K, Elleuch F (2005) Daily and monthly averaged aerosol optical properties and diurnal variability deduced from AERONET sun-photometric measurements at Thala site (Tunisia). Meteorol Atmos Phys 92:103–114. doi:10.1007/s00703-005-0137-8

Emmerechts J, Jacobs L, Hoylaerts MF (2011) Air pollution and cardiovascular disease. In: Khallaf MK (ed) Impact air pollution health, economy, environment and agricultural sources. InTech, pp 69–92

Ghan SJ, Schwartz SE (2007) Aerosol properties and processes: a path from field and laboratory measurements to global climate models. Bull Am Meteorol Soc 88:1059–1083. doi:10.1175/BAMS-88-7-1059

Gueymard CA (2008) Prediction and validation of cloudless shortwave solar spectra incident on horizontal, tilted, or tracking surfaces. Sol Energy 82:260–271. doi:10.1016/j.solener.2007.04.007

Holben BN, Eck TF, Slutske I et al (1998) AERONET–A federated instrument network and data archive for aerosol characterization. Remote Sens Environ 66:1–16

Huang H (2009) Aerosol remote sensing using AATSR. University of Oxford, Oxford

Kiedron PW, Michalsky JJ, Berndt JL, Harrison LC (1999) Comparison of spectral irradiance standards used to calibrate shortwave radiometers and spectroradiometers. Appl Opt 38:2432–2439

Kiehl JT, Schneider TL, Rasch PJ et al (2000) Radiative forcing due to sulfate aerosols from simulations with the National Center for Atmospheric Research Community Climate Model, Version 3. J Geophys Res 105:1441. doi:10.1029/1999JD900495

Lohmann U (2006) Aerosol effects on clouds and climate. Space Sci Rev 125:129–137. doi:10.1007/s11214-006-9051-8

Mishchenko MI, Geogdzhayev IV, Cairns B et al (2007) Past, present, and future of global aerosol climatologies derived from satellite observations: a perspective. J Quant Spectrosc Radiat Transf 106:325–347

Myhre G, Berglen TF, Johnsrud M et al (2009) Modelled radiative forcing of the direct aerosol effect with multi-observation evaluation. Atmos Chem Phys 9:1365–1392

Nair SK, Sijikumar S, Prijith SS (2012) Impact of continental meteorology and atmospheric circulation in the modulation of aerosol optical depth over the Arabian Sea. J Earth Syst Sci 121:263–272

Nieke J, Pflug B, Zimmermann G (1999) An aureole-corrected Langley-plot method developed for the calibration of HiRES grating spectrometers. J Atmos Solar-Terrestrial Phys 61:739–744

Paul R, Jagdish G, Kuniyal C et al (2011) The assessment of aerosol optical properties over Mohal in the Northwestern Indian Himalayas using satellite and ground-based measurements and an influence of aerosol transport on aerosol radiative forcing. Meteorol Atmos Phys 153–169. doi:10.1007/s00703-011-0149-5

Pöschl U (2005) Atmospheric aerosols: composition, transformation, climate and health effects. Angew Chem Int Ed Engl 44:7520–7540. doi:10.1002/anie.200501122

Remer LA, Kaufman YJ, Tanre D et al (2005) The MODIS aerosol algorithm, products, and validation. J Atmos Sci 62:947–973

Saeed TM, Al-Dashti H (2010) Optical and physical characterization of "Iraqi freedom" dust storm, a case study. Theor Appl Climatol 104:123–137. doi:10.1007/s00704-010-0334-3

Satheesh SK, Krishna Moorthy K, Kaufman YJ, Takemura T (2005) Aerosol optical depth, physical properties and radiative forcing over the Arabian Sea. Meteorol Atmos Phys 91:45–62. doi:10.1007/s00703-004-0097-4

Schmid B, Wehrli C (1994) High precision of calibration of a sun photometer using Langley plots performed at Jungfraujoch (3580 m) and standard irradiance lamps. Geosci. Remote Sens. Symp. 1994. IGARSS'94. Surf. Atmos. Remote Sens. Technol. Data Anal. Interpret. Int. pp 2314–2316

Shaw GE (1983) Sunphotometry. Bull Am Meteorol Soc 64:4–10

Slusser J, Gibson J, David B et al (2000) Langley method for calibrating UV filter radiometer. J Geophys Res 105:4841–4849

Twomey S (1974) Pollution and the planetary albedo. Atmos Environ 8:1251–1256

Whitby KT, Husar RB, Liu BYH (1972) The aerosol size distribution of Los Angeles smog. J Colloid Interface Sci 39:177–204

Yu H, Kaufman YJ, Chin M et al (2006) A review of measurement-based assessments of the aerosol direct radiative effect and forcing. Atmos Chem Phys 6:613–666. doi:10.5194/acp-6-613-2006

Chapter 2
Ground-Based Aerosol Optical Depth Measurements

Abstract This chapter presents a detailed overview on the theory of aerosol optical depth retrieval with the emphasis on ground-based sunphotometry technique. To further discuss on the calibration issue as mentioned previously, this overview also includes the principle of the oldest passive ground-based calibration method, Langley calibration to provide an insight on the working mechanism of the method. The final part in this chapter compiles the previous existing Langley calibration method in a chronological sequence to render a better comprehension on its development from the past to present time.

Keywords Aerosol optical depth · Sunphotometers · Spectroradiometer · Langley calibration

2.1 Theory of Aerosol Absorption and Scattering

The fundamental theory of aerosol absorption and scattering is explained by Mie theory which describes the attenuation of electromagnetic radiation by spherical particles through solving the Maxwell equations. Mie theory is also called Lorenz-Mie theory or Lorenz-Mie-Debye theory. In this theory, there are two important key assumptions: (1) particle is a sphere, and (2) particle is homogenous and therefore it is characterized by single refractive index $m = n - ik$ at a given wavelength. Mie theory requires the relative refractive index that is the refractive index of a particle divided the refractive index of a medium. In this case, the medium of atmospheric aerosol is often assumed in air of m is about 1 and since it has complex chemical composition, the effective refractive index is often calculated at a given wavelength.

Basically, Mie theory calculates the scattered electromagnetic field at all points within the particle which called internal field and at all points of the homogeneous medium in which the particle is embedded as shown in Fig. 2.1. In almost all practical applications in atmosphere, light scattering observations are carried out in

J. Dayou et al., *Ground-Based Aerosol Optical Depth Measurement Using Sunphotometers*, SpringerBriefs in Applied Sciences and Technology, DOI: 10.1007/978-981-287-101-5_2, © The Author(s) 2014

Fig. 2.1 Simplified
visualization of scattering of
an incident EM wave by
particle

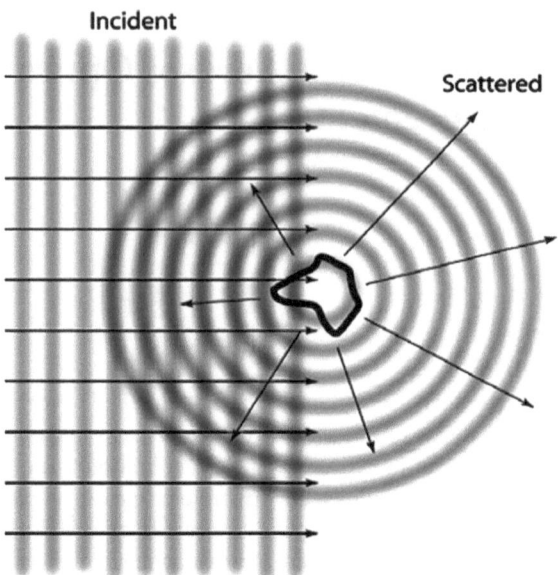

the far-field zone (i.e., at the large distances R from a particle). The solution of the
wave equation in the far field zone can be obtained as (Mie 1908)

$$\begin{bmatrix} E_l^s \\ E_r^s \end{bmatrix} = \frac{\exp(-ikR + ikI)}{ikR} \begin{bmatrix} S_2(\theta) & S_3(\theta) \\ S_4(\theta) & S_1(\theta) \end{bmatrix} \begin{bmatrix} E_l^i \\ E_r^i \end{bmatrix}, \tag{2.1}$$

where $k = 2\pi/\lambda$, E_l^i and E_r^i are the parallel and perpendicular components of
incident electrical field, and E_l^s and E_r^s are the parallel and perpendicular compo-
nents of scattered electrical field, $\begin{bmatrix} S_2(\theta) & S_3(\theta) \\ S_4(\theta) & S_1(\theta) \end{bmatrix}$ is the amplitude scattering
matrix at scattering angle θ.

As the first assumption suggested, for spherical particle, $S_3(\theta)$ and $S_4(\theta)$ are
zero, and thus Eq. 2.1 gives the fundamental equation of scattered radiation by a
sphere including polarization as

$$\begin{bmatrix} E_l^s \\ E_r^s \end{bmatrix} = \frac{\exp(-ikR + ikI)}{ikR} \begin{bmatrix} S_2(\theta) & 0 \\ 0 & S_1(\theta) \end{bmatrix} \begin{bmatrix} E_l^i \\ E_r^i \end{bmatrix}. \tag{2.2}$$

In Eq. 2.2, the Mie theory defines scattering amplitudes $S_1(\theta)$ and $S_2(\theta)$ function
as (Mie 1908)

$$S_1(\theta) = \sum_{n=1}^{\infty} \frac{2n+1}{n(n+1)} [a_n \pi_n(\cos\theta) + b_n \tau_n(\cos\theta)], \tag{2.3}$$

$$S_2(\theta) = \sum_{n=1}^{\infty} \frac{2n+1}{n(n+1)} [b_n \pi_n(\cos\theta) + a_n \tau_n(\cos\theta)], \qquad (2.4)$$

where $\pi_n(\cos\theta)$ and $\tau_n(\cos\theta)$ are Mie angular functions written as

$$\pi_n(\cos\theta) = \frac{1}{\sin(\theta)} P_n^1(\cos\theta), \qquad (2.5)$$

$$\tau_n(\cos\theta) = \frac{d}{d\theta} P_n^1(\cos\theta), \qquad (2.6)$$

where P_n^1 are the associated Legendre polynomials, a_n and b_n are scattering coefficient in the function of size parameter x.

To determine the scattering phase function P_n^1 (cos θ), Mie theory relates the Stoke parameters $\{I_o, Q_o, V_o, U_o\}$ of incident radiation field and Stoke parameters $\{I, Q, V, U\}$ of scattered radiation as (Bohren and Huffman 1983)

$$\begin{bmatrix} I \\ Q \\ V \\ U \end{bmatrix} = \frac{\sigma_s}{4\pi r^2} P \begin{bmatrix} I_o \\ Q_o \\ V_o \\ U_o \end{bmatrix}, \qquad (2.7)$$

where P is defined as

$$P = \begin{bmatrix} P_{11} & P_{12} & 0 & 0 \\ P_{12} & P_{22} & 0 & 0 \\ 0 & 0 & P_{33} & -P_{34} \\ 0 & 0 & P_{34} & P_{44} \end{bmatrix}. \qquad (2.8)$$

In a particle of any shape, the scattering phase function consists of 16 independent elements, but for a spherical particle this number reduces to four as $P_{22} = P_{11}, P_{44} = P_{33}$. Thus, for sphere, Eq. 2.7 is rewritten as

$$\begin{bmatrix} I \\ Q \\ V \\ U \end{bmatrix} = \frac{\sigma_s}{4\pi r^2} \begin{bmatrix} P_{11} & P_{12} & 0 & 0 \\ P_{12} & P_{11} & 0 & 0 \\ 0 & 0 & P_{33} & -P_{34} \\ 0 & 0 & P_{34} & P_{33} \end{bmatrix} \begin{bmatrix} I_o \\ Q_o \\ V_o \\ U_o \end{bmatrix}. \qquad (2.9)$$

where σ_s and r are the scattering cross-section and radius of the particle respectively.

From Mie theory, it is given that the extinction cross-section of a particle with radius r in the forward direction $\theta = 0°$ as (Bohren and Huffman 1983)

$$\sigma_s = \frac{4\pi}{k^2} \text{Re}[S(\theta = 0°)]. \tag{2.10}$$

Following Eqs. 2.3 and 2.4 in $\theta = 0°$, both equations yield

$$S_1(0°) = S_2(0°) = \frac{1}{2} \sum_{n=1}^{\infty} (2n+1)(a_n + b_n). \tag{2.11}$$

The efficiencies Q of extinction, scattering, absorption are defined as

$$Q = \frac{\sigma}{\pi r^2}, \tag{2.12}$$

where σ is the cross-section of a particle with radius r. This brings the solution for Q_e, Q_s, and Q_a in terms coefficents a_n and b_n as

$$Q_e = \frac{2}{x^2} \sum_{n=1}^{\infty} (2n+1)\text{Re}(a_n + b_n), \tag{2.13}$$

$$Q_s = \frac{2}{x^2} \sum_{n=1}^{\infty} (2n+1)(|a_n|^2 + |b_n|^2), \tag{2.14}$$

$$Q_a = Q_e - Q_s, \tag{2.15}$$

where x is the size parameter.

2.1.1 The Optical Properties of Spherical Particle

In the previous section, Mie theory was used to define the extinction, scattering, and absorption cross-section as a function of particle size and wavelength. In this section, the cross section of extinction, scattering, and absorption is integrated over the size distribution $N(r)$ to yield the optical properties of an ensemble spherical particle.

For a given type of particles characterized by the size distribution $N(r)dr$, the volume extinction K_e, scattering K_s and absorption K_a coefficients are determined as (King et al. 1978)

$$K_e = \int_{r_1}^{r_2} \sigma_e(r)N(r)dr = \int_{r_1}^{r_2} \pi r^2 Q_e N(r)dr; \tag{2.16}$$

$$K_s = \int_{r_1}^{r_2} \sigma_s(r)N(r)dr = \int_{r_1}^{r_2} \pi r^2 Q_s N(r)dr; \tag{2.17}$$

Fig. 2.2 Intrinsic
visualisation of transmission
of an extraterrestrial radiation
$I_{\lambda,o}$ through an optical path
length s_1 and s_2

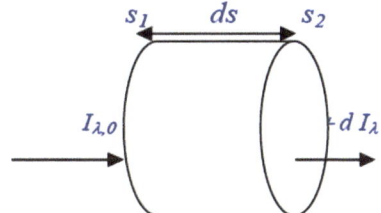

$$K_a = \int_{r_1}^{r_2} \sigma_a(r)N(r)dr = \int_{r_1}^{r_2} \pi r^2 Q_a N(r)dr. \tag{2.18}$$

On the other hand, for external mixture that contains several types of particles, the total effective volume extinction, scattering and absorption are the sum of each particle component as

$$K_e = \sum_i K_e^i = \sum_i \int_{r_1}^{r_2} \sigma_e(r)N_i(r)dr = \sum_i \int_{r_1}^{r_2} \pi r^2 Q_e N_i(r)dr, \tag{2.19}$$

$$K_s = \sum_i K_s^i = \sum_i \int_{r_1}^{r_2} \sigma_s(r)N_i(r)dr = \sum_i \int_{r_1}^{r_2} \pi r^2 Q_s N_i(r)dr, \tag{2.20}$$

$$K_a = \sum_i K_a^i = \sum_i \int_{r_1}^{r_2} \sigma_a(r)N_i(r)dr = \sum_i \int_{r_1}^{r_2} \pi r^2 Q_a N_i(r)dr, \tag{2.21}$$

where K_e^i, K_s^i, and K_a^i are calculated for each particle type characterized by its particle size distribution $N_i(r)$ and a refractive index m_i.

On the whole, the fundamental theory of Mie particle scattering and absorption resulted in the main radiation law of extinction which also known as Beer-Lambert-Bouger's Law. It states that the extinction process is linear in the intensity of radiation and amount of matter, provided that the physical state (i.e., temperature, pressure, composition) is held constant. Consider a small volume ΔV of infinitesimal length ds and unit area ΔA containing optically active matter (gases, aerosols, and/or cloud drops) as shown in Fig. 2.2. The change of intensity along a path ds of a light with wavelength λ is proportional to the amount of matter in the path as

$$dI_\lambda = -K_{e,\lambda}I_{\lambda,o}ds, \tag{2.22}$$

where $K_{e,\lambda}$ is the volume extinction coefficient obtained from Eq. 2.19 and, $I_{\lambda,o}$ is the source of function which in this case is the extraterrestrial solar radiation at zero air mass. The optical depth of a specific layer between s_1 and s_2 is then determined by integrating Eq. 2.22 in optical path s as (Dubovik and King 2000)

$$\tau_{\lambda(s_1,s_2)} = \int_{S_1}^{S_2} K_{e,\lambda} ds. \tag{2.23}$$

Finally, it leads to the well-known Beer-Lambert-Bouger's Law of extinction as

$$I_\lambda = I_{\lambda,o} \exp\left(\int_{S_1}^{S_2} -K_{e,\lambda} ds\right) = I_{\lambda,o} \exp(-\tau_\lambda). \tag{2.24}$$

2.2 Ground-Based Aerosol Optical Depth Retrieval

The two most well-known ground-based AOD retrieval techniques are sunphotometry and lidar. The former is a passive optical system that measures the extinction of direct-beam radiation in distinct wavelengths, and retrieves the aerosol contribution to the total extinction. The latter is an active optical system transmits light into the atmosphere and then collects the backscatter light signals to retrieve the aerosol attenuation in total columnar atmosphere. Details of each technique are separately discussed in the following sub-sections.

2.2.1 Retrieval with Ground-Based Sunphotometry Radiometer

Unlike the satellite data that uses upwelling radiances viewed from space, ground-based sunphotometry radiometer uses the down-welling radiances of solar radiation to retrieve total columnar AOD in a specific area. Under cloudless condition, the higher the extinction value in the solar transmission corresponds to higher aerosol loading. Figure 2.3 demonstrates this effect for increasing AOD from 0 to 1 at constant air mass (AM1.5). The blue line at the top represents the extraterrestrial solar spectrum at zero air mass (AM0), which is considered as aerosol-free spectrum. The solar spectrum in direct normal irradiance (DNI) gradually decreases for increasing AOD due to scattering and absorption caused atmospheric aerosols for all wavelengths. From the figure, it is obvious that higher attenuation is experienced by light in the mid-visible range of the solar spectrum compared to other parts of the spectrum. This explains why in most aerosol measurements, visible range wavelengths are regularly used for spectral AOD retrieval. To provide an overview of the ground-based sunphotometry measurement, a number of selected aerosol monitoring networks including Aerosol Robotic Network (AERONET), Multi-Filter Rotating Shadowband Radiometer (MFRSR), China Meteorological Administration Aerosol Remote Sensing Network (CASRNET), and Maritime Aerosol Network (MAN) are briefly discussed as follows.

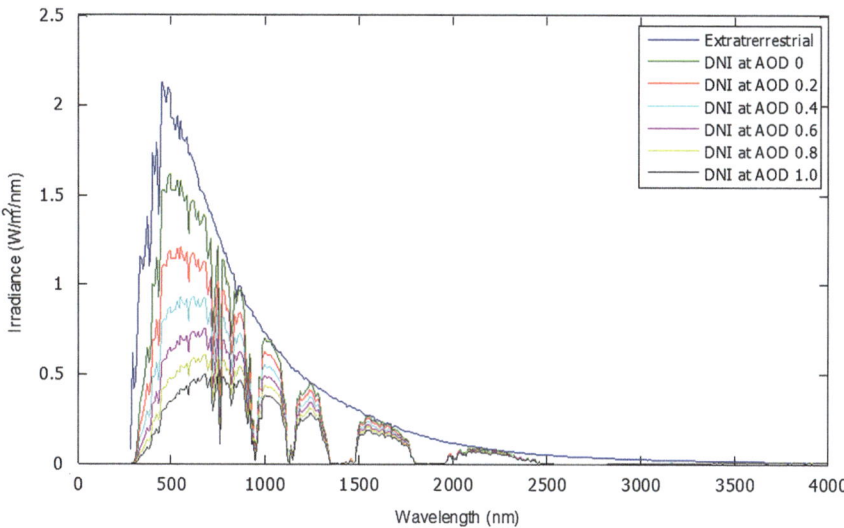

Fig. 2.3 Diminution of solar transmission at multiple AOD values from 0 to 1. Simulation is based on urban aerosol model over tropical atmosphere

2.2.1.1 Aerosol Robotic Network (AERONET)

At global scale over land, AOD is monitored by AERONET project, a federation of ground-based remote sensing aerosol networks established by National Aeronautics and Space Administration (NASA) and PHysics, Optoelectronics, and Technology of Novel Micro-resonator Structures (PHOTONS). It is greatly expanded by collaborators from national agencies, institutes, universities, and individual scientist with currently over 120 monitoring stations across the world, but this does not provide global coverage as illustrated in the map shown in Fig. 2.4. It uses sun photometer CIMEL CE-318 to retrieve AOD within the spectral range 340–1020 nm by means a filtered detector that measures the spectral extinction of direct beam radiation according to Beer-Lambert-Bouguer's Law.

Approximately every 15 min, the sunphotometer points directly at the sun during the daytime, taking spectral measurements in triplicate over 1.5 min. Cloud-screening algorithm is performed by limiting the variability within each triplet and compared to prior and subsequent triplets (Smirnov et al. 2000). Estimates of Angstrom's exponent, and separation into fine and coarse mode contributions, can also be computed via the spectral de-convolution algorithm of O'Neill et al. (2001). Despite the direct sun measurements, AERONET instruments are also programmed to observe angular distribution of sky radiance in approximately every hour during the daytime. These sky measurements are used to retrieve size distribution and scattering/extinction properties of the ambient aerosol using spherical aerosol assumptions (Dubovik and King 2000). By assuming the ambient aerosol to be polydisperse spheres and randomly oriented spheroids, the algorithm

 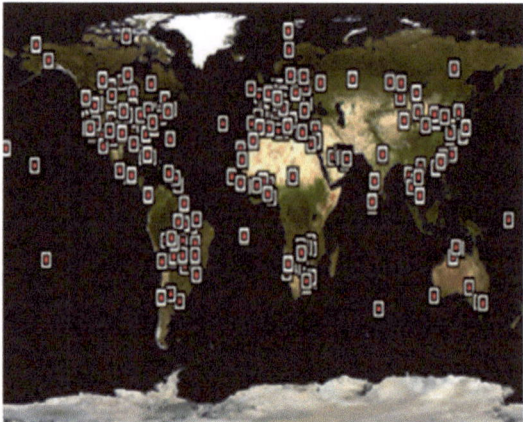

Fig. 2.4 AERONET networks worldwide and CIMEL sunphotometer—adapted from AERONET NASA in http://aeronet.gsfc.nasa.gov/

retrieves the volume distribution that corresponds to the best fit of both sun-measured AOD and sky radiances.

Retrievals from both sun and sky AERONET measurements are controlled by rigorous calibration and cloud-screening algorithms. Limitations of the retrieval also relate to low optical depth conditions, angular coverage of sky radiance measurements, and azimuth angle pointing of the instrument (Dubovik et al. 2000). Nevertheless, it is expected that precise aerosol characterization in the absorption and optical properties could yield accurate retrieval results that can be used as ground-truth estimates (Dubovik et al. 2002).

2.2.1.2 Multifilter Rotating Shadowband Radiometer Network

Ground-based aerosol monitoring network is also supplemented by multifilter rotating shadowband radiometer (MFRSR) networks which measure total and diffuse solar irradiances at multiple wavelengths using shadowband technique. Instead of using the narrow field-of-view (FOV) approximation, direct solar irradiance is obtained by subtracting the diffuse radiation from the total irradiance. This new robotic instruments are mostly operated in remote areas over United States and usually unattended (Augustine et al. 2003). Thus, their raw data represent wide range of atmospheric condition, which are undesirable for AOD retrieval, especially when clouds obscure the sun. Currently, the network uses independent cloud screening algorithm to filter the data suitable for spectral AOD retrieval (Alexandrov et al. 2004). The proposed algorithm characterizes the degree of horizontal inhomogeneity of an atmospheric field. It provides computational efficiency and the ability to detect short clear sky intervals under broken cloud cover conditions.

Furthermore, most MFRSRs are not calibrated against standard references but calibrated in a relative sense from their own operational data (Augustine et al. 2003). The usual calibration method used is Langley plot technique, in which the instrument's output at TOA is inferred by extrapolating to zero air mass. Once a stable extrapolated value is obtained, it can be used for AOD retrieval within the calibration period.

2.2.1.3   China Meteorological Administration Aerosol Remote Sensing Network

In China, it has its own aerosol monitoring program called China Meteorological Administration Aerosol Remote Sensing Network (CARSNET), established in 2002 (Che et al. 2009). It is a routine operation network, purposely launched for the study of aerosol properties and for validation of satellite aerosol retrievals. Similar to AERONET, it deploys the CIMEL sunphotometer for the measurement of direct spectral solar radiance. The CARSNET sunphotometer is annually calibrated to ensure its performance and quality of measurements.

CARSNET has established an independent calibration system that is consistent to AERONET (Che et al. 2009). It uses two master instruments to inter-calibrate all other instruments of the network. The two masters are calibrated by the Langley method following the AERONET protocols in Izana Observatory. In order to contain the uncertainty caused by the degradation of the instruments themselves, the two master instruments of CARSNET are calibrated periodically every three or six months at Waliguan Mountain (36 °17'N, 100 °55'E, 3816 m) the Global Atmosphere Watch (GAW) station of China alternately.

2.2.1.4   Maritime Aerosol Network

To extend the aerosol monitoring over ocean, Maritime Aerosol Network (MAN) as a component of AERONET has been established since November 2006. MAN employs the Microtops handheld sunphotometer and utilizes calibration and data processing procedures traceable to AERONET. A valid comparison among various models, satellite products and sunphotometer measurement suggested that majority of the AOD differences are positive by a factor of 0.2 at most (Smirnov et al. 2011). The discrepancy is believed to be, at least partly, by uncertainties in aerosol production rates, foam formation and its latitudinal distribution, cloud contamination, accuracy of radiative transfer model used, surface reflectance effects.

2.3 Conventional Langley Calibration Method

This section explains the principle of the oldest passive ground-based calibration method, Langley calibration. Until now, the stated method is widely used in many AOD monitoring networks because it requires no additional calibration equipments. However, its accuracy is strongly governed by the atmospheric condition where the calibration is performed. For an ideal condition, it is often performed at high altitude to avoid abundant cloud cover and high unstable aerosol content.

The Langley method uses the changes of observed path length through the atmosphere to compute an optical depth (Harrison and Michalsky 1994). It is a method to measure the sun's irradiances with ground-based instruments that based on repeated measurements operated at a given location for a cloudless morning or afternoon, as the sun moves across the sky for significant changes of air mass or path length. A successful Langley plot is imperative to permit extrapolation of the regression line to air mass zero. The extrapolated value further allows the determination of the instrument output at top of atmosphere or better known as extraterrestrial value. This value is then useful in radiometric calibration when divided by spectrally extraterrestrial irradiances constant (Nieke et al. 1999; Schmid and Wherli 1994). It can also be used in aerosol optical depth retrieval when divided with solar irradiances measured at ground after subtractions by other relevant optical depths.

Figure 2.5 presents an idealized Langley plot at 500 nm for different AOD values from 0.02 to 0.30. It should be noted that no specific reason is inherited in the negative values of irradiance in the y-axis, as value of <1 produces negative natural logarithm. From the figure, it clearly shows that extrapolation to zero air mass by Langley plot for AOD 0.02–0.10 gives a nearly consistent extraterrestrial constant, though with some negligible errors. However, when under high aerosol content (refer line AOD 0.20 and 0.30 in Fig. 2.5), the extrapolation could incur serious inaccuracy even for highly stable aerosol content. Therefore, this leads to believe that for an ideal performance of Langley plot at any wavelengths, a highly stable and low in magnitude aerosol loading is of necessary important.

Its working principle lies on the basis that as the solar radiation transmits through atmosphere, it experiences a stream of attenuations either by absorption or scattering due to the air molecules or solid particles suspended in the atmosphere. From Eq. 2.24, the attenuation of an electromagnetic radiation through an optical path length can be described by the Beer-Lambert-Bougeur's Law. By applying the fundamental theory of Mie scattering and absorption in Sun's direct-beam monochromatic radiation passing through the Earth's atmosphere, it obeys the extinction law of exponential as (Thomason et al. 1983)

$$I_\lambda = R^2 I_{o,\lambda} \exp\left(-\sum \tau_{\lambda,i} m_i\right) \tag{2.25}$$

where I_λ is the direct normal irradiance at the ground at wavelength λ, R is the Earth-to-Sun distance in astronomical units (AU), $I_{o,\lambda}$ is the extraterrestrial

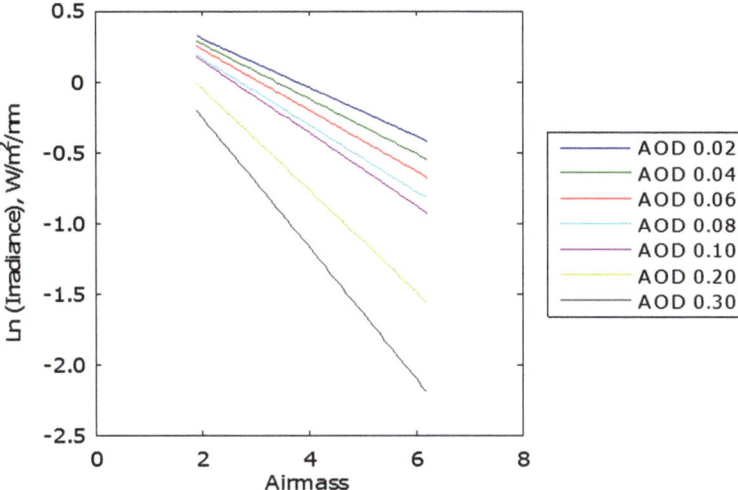

Fig. 2.5 Idealized Langley plot at 500 nm for multiple AOD values. Simulation is based on urban aerosol model over tropical atmosphere

irradiance at the top of atmosphere (TOA), $\tau_{\lambda,i}$ is the total optical depth of the ith scatterer or absorber, and m_i is the air mass of the ith scatterer or absorber through the atmosphere. Taking the natural logarithm of both sides, Eq. 2.25 can be written as

$$\ln I_\lambda = \ln R^2 I_{o,\lambda} - \sum \tau_{\lambda,i} m_i. \tag{2.26}$$

The total optical depth, $\tau_{\lambda,i}$ in Eq. 2.26 is contributed by Rayleigh, ozone, aerosol and trace gases, which can be written as

$$\tau_{\lambda,i} = \tau_{R,i} + \tau_{o,i} + \tau_{a,i} + \tau_{g,i}. \tag{2.27}$$

The Rayleigh contribution is approximated using the relationship (Djamila et al. 2011; Knobelspiesse et al. 2004)

$$\tau_{R,\lambda,i} \propto \frac{p}{p_o} \exp\left(-\frac{H}{7998.9}\right), \tag{2.28}$$

where p is the site's atmospheric pressure, p_o is the mean atmospheric pressure at sea-level and H is the altitude from sea-level in meter. Meanwhile, the ozone optical depth can be estimated through the satellite observation of ozone concentration, C_o in Dobson unit (DU) (Knobelspiesse et al. 2004)

$$\tau_{o,\lambda,i} \propto C_o. \tag{2.29}$$

Other major trace gases contributions are nitrogen dioxide and sulphur dioxide. These contributions are more dominant in highly urbanized or industrial area. Similarly, their optical depth can be estimated through the satellite observation or ground-based measurement of their respective concentration.

According to Eq. 2.26, the uncalibrated pixels (counts, P) measured by the spectrometer is then

$$\ln P_\lambda = \ln R^2 P_{o,\lambda} - \sum \tau_{\lambda,i} m_i, \tag{2.30}$$

where $P_{o,\lambda}$ is the extrapolated pixels intercept at zero air mass. When the range of interest is restricted in visible bands, the trace gases contributions can be neglected. In this way, the remaining contributions are now constrained to Rayleigh and ozone only. Substituting Eq. 2.27 into 2.30 gives the final equation as

$$\ln P_\lambda + \tau_{R,i} m_i + \tau_{o,i} m_i = \ln R^2 P_{o,\lambda} - \tau_{a,i} m_i. \tag{2.31}$$

Using this approximation, changes in Rayleigh optical depth due to pressure fluctuation and nominal ozone optical depth at each point is subtracted. Thus, leaving the left side of Eq. 2.31 insensitive to pressure variations and differences caused by ozone (Michalsky and Kiedron 2008). This step is important because small changes in pressure or ozone column during the observation can noticeably affect the extrapolated values.

On a clear day, a Langley plot gives a stable $P_{o,\lambda}$ for each wavelength when the data are extrapolated to TOA. With sufficient data of $P_{o,\lambda}$ on several clear days, an averaged $P_{o,\lambda(avg)}$ can be obtained using the following equation

$$P_{o,\lambda(avg)} = \frac{1}{n} \sum_{i=n}^{i} P_{o,\lambda,n}, \tag{2.32}$$

n is the number of Langley plots available for calibration. Accordingly, the calibration factor k, is obtained by dividing the averaged extrapolated values with the extraterrestrial constant. Finally, the calibrated irradiance measured by the spectrometer is determined by multiplying the pixels measured at the ground P_λ with the calibration factor k as (Slusser et al. 2000)

$$I_\lambda = P_\lambda k = \frac{P_\lambda \int P_{o,\lambda(avg)} F_\lambda d\lambda}{I_{o,\lambda} \int F_\lambda d\lambda}. \tag{2.33}$$

2.4   Historical Development of Langley Calibration Method

The Langley calibration method can be categorized into five types chronologically. The oldest method is the basic sunphotometry Langley method. This is then followed by circumsolar Langley method, cloud-screened Langley method, maximum value composite (MVC) Langley method, and comparative Langley method. Details of each method are elaborated in the following.

2.4.1   Basic Sunphotometry Langley Method

The diminution of light passing through the atmosphere was for the first time quantified by Pierre Bouguer in 1725 (Shaw 1983). This attenuation was found increasing exponentially to the evolution of optical path length and led to the establishment of the well known Beer-Lambert-Bouguer's principle of exponential. For a thin layer of atmosphere, it could be considered as planar and therefore the passage of light through a pane of colored glass could be used to explain the mechanism of the transmission. In both mechanism, the optical transmission T obeys

$$T = \exp(\frac{-\varphi l}{\cos z}), \tag{2.34}$$

where l is the thickness of the medium, z is the angle of the beam of primary illumination, and ϕ is the optical index of the medium, which in this case is the turbidity optical index.

The determination of T in Eq. 2.34 is difficult as the primary incident illumination is inaccessible. Therefore, Eq. 2.34 should be considered in terms of an arbitrary reference at two angles of primary illumination Z_1, and Z_2, then the ratio is given as

$$\ln(I_1/I_2) = \beta l[1/\cos z_1 - 1/\cos z_2], \tag{2.35}$$

where I represents the solar or lunar light intensity which is independent of I_o (the extraterrestrial solar light intensity) and the quantity βl is the quantity to be determined. In the case of glass plate, l is the thickness of the medium. However, in the atmosphere it represents the total optical path length travelled by I. This principle when used in Langley plot can be used to determine the instrument's output at TOA, which is useful in calibration for AOD retrieval. However, one important task in Langley calibration is to insure the temporal drifts in atmospheric transmissivity do not lead to erroneous calibration constants (Shaw 1983). The only way to achieve this is to perform the Langley calibration from an excellent high altitude mountain observatory.

2.4.2 Circumsolar Langley Method

Not long after that, a modified Langley calibration was developed, in which simultaneous measurements of circumsolar radiation are incorporated (Tanaka et al. 1986). In this method, the logarithm of the sunphotometer reading is plotted against the ratio of intensity of singly scattered circumsolar radiation to that of direct solar radiation instead of the optical air mass as in the conventional Langley method. The idea of using circumsolar radiation for monitoring atmospheric turbidity is based on the availability of the data of circumsolar radiation for quantitative detection of very small amounts of aerosols and other particulates and of small changes in their concentration, size and composition.

The single-scattering approximation of the circumsolar (aureole) intensity in the almucantar of the sun is given by

$$F_a^1(\mu_o, \varphi) = m\tau\omega_o P(\cos\theta)F_o \exp(-m\tau)\Delta\Omega, \tag{2.36}$$

where μ_o is the cosine of the solar zenith angle, ϕ is the azimuthal angle measured from the solar principal plane, ω_o is the single scattering albedo, $P(cos\theta)$ is the normalized phase function at the scattering angle θ, $\Delta\Omega$ is the solid viewing angle of the radiometer, and $cos\theta$ is given by

$$\cos\theta = \mu_0^2 + (1 - \mu_0^2)\cos\varphi. \tag{2.37}$$

From Eq. 2.36, the intensity of singly scattered radiation in the solar almucantar is proportional to the optical depth m_T contributed by aerosols and air molecules given by

$$\tau_T = \tau_a + \tau_m, \tag{2.38}$$

$$\omega_o = (\omega_{oa}\tau_a + \omega_{om}\tau_m)/\tau, \tag{2.39}$$

$$P(\cos\theta) = [\omega_{oa}\tau_a P_a(\cos\theta) + \omega_{om}\tau_m P_m(\cos\theta)]/\omega_o\tau, \tag{2.40}$$

where τ_a, ω_{oa}, and $P_a(cos\theta)$ are the optical depth, the single scattering albedo and the phase function of aerosols, respectively; and τ_m, ω_{om}, and $P_m(cos\theta)$ are corresponding quantities for air molecules. The phase function $P(cos\theta)$ is defined to satisfy the normalization integral of

$$2\pi \int_{-1}^{1} P(\cos\theta)d\theta = 1. \tag{2.41}$$

When simultaneous measurement of the intensity of direct solar radiation and that of circumsolar radiation from a given portion of the aureole region is made by a single radiometer, Eqs. 2.36–2.40 can be combined to form

$$\ln F = \ln F_o - \tau^*, \tag{2.42}$$

$$\tau^* = m\tau = F_a^1/[F\Delta\Omega\omega_o P(\cos\theta)]. \tag{2.43}$$

Given that the magnitudes of $P(cos\theta)$ are more or less independent of the size distribution of aerosols at scattering angle around 20°, the intensity of singly scattered radiation F_a^1 from the measured intensity F_a by can be determined as

$$F_a^1 = SSR(m, \tau_a, \tau_m, \overline{m}, \omega, \theta)F_a, \tag{2.44}$$

where *SSR* is the single scattering ratio that depends on several parameters such as optical air mass m, optical depth of aerosols τ_a and air molecules τ_m, complex index of refraction of aerosol $\overline{m}$, ground albedo ω, and scattering angle θ.

Figure 2.6 presents an example of the modified Langley plot for turbidity condition of $\tau_a = 0.2$ at noon. The three regression lines in Langley plot on the left panel are simulated based on Shaw's parabolic drift parameter assumed α to be 0, 0.011 and -0.011, corresponds to change of τ_a in 0, 10, and -10 % for 3 h around noon. It is evident from the figure that the Langley-plot method predicts the symmetrically larger or smaller values of F_o for finite values of α despite of an excellent linearity in the respective plots. The circumsolar Langley method on the right panel improved the consistency and accuracy in 5–10 times for the wavelengths greater than 500 nm in spite of varying τ_a. Measurement of circumsolar radiation for the aureole-corrected Langley calibration had also been realized by pointing the observation direction to the side of the Sun instead of using theoretical circular aureole ring measurements (Nieke et al. 1999). Similar results were reported that inclusion of the aureole signal measurements significantly reduced the deviation compared to the classical Langley-plot analysis.

2.4.3 Cloud-Screened Langley Method

A sunphotometry Langley calibration particularly for the large network of automated instrument which collects large pool of data under both cloudless and noncloudless condition needs to be cloud-screened prior to the calibration. For instance, the MFRSRs are usually operated in remote areas and unattended. Thus, their raw data represent a wide range of atmospheric conditions which may undesirable for AOD analysis when clouds obscure the sun. Moreover, MFRSRs are typically not calibrated against standard references and therefore must be calibrated in a relative way from their own measurement data (Augustine et al. 2003). Besides, a good cloud-screening algorithm should work on raw un-calibrated data as the use of the measurement's spectral signature for cloud screening can be affected by initial calibration uncertainties and thus results in ineffective cloud-screening. Owing to these issues, the Langley calibration method continued

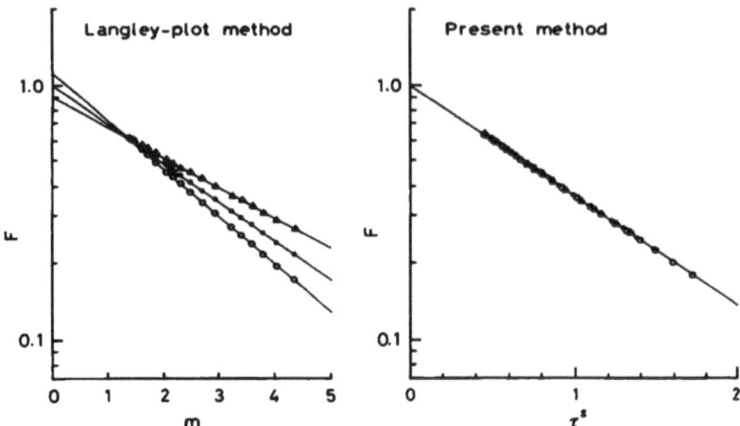

Fig. 2.6 Comparison between the Langley-plot method (*left panel*) and the Circumsolar Langley method (*right panel*) at $\lambda = 500$ nm (Tanaka et al. 1986)

to improve for an objective cloud-screening algorithm was introduced to select appropriate data from a continuous time series that needed for the regression.

2.4.3.1 Statistical Filters

The pioneer in this development was operating on a time series of direct normal irradiance observations, which can be described as a series of sequential filters that rejects undesirable points (Harrison and Michalsky 1994). In this method, the first filter is a forward finite-different derivative filter that identifies regions where the slope of Langley plot is positive. These cannot be produced by any uniform air mass turbidity process and are evidence of the recovery of the direct-normal irradiance from a cloud transits. The second filter is a subsequent finite-difference derivative filter tests for regions of strong second derivatives. In this case, regions that are more than twice the mean are eliminated. In other words, this filter rejects points near the edge of intervals eliminated by the first filter, if it was insufficiently aggressive, and also eliminates any cloud passage that occurs at the end of the sampling interval. To further affect a robust linear regression, two iterations are imposed by performing a conventional least-square regression on the remaining points from each filtration and a sweep is made through data points that have more than 1.5 standard deviation from the regression line or residual less than 0.006.

An example that demonstrates the Langley regression for the morning interval identified by the objective algorithm is presented in Fig. 2.7. In the figure, the points marked with "+" contribute equally to the regression and points marked with small open box were removed by the derivative filters in the time-series plot near AM = 2.0. The first iteration removes a weak cloud passage in the time-series plot around AM = 4.5. These points are marked with a box and cross. The

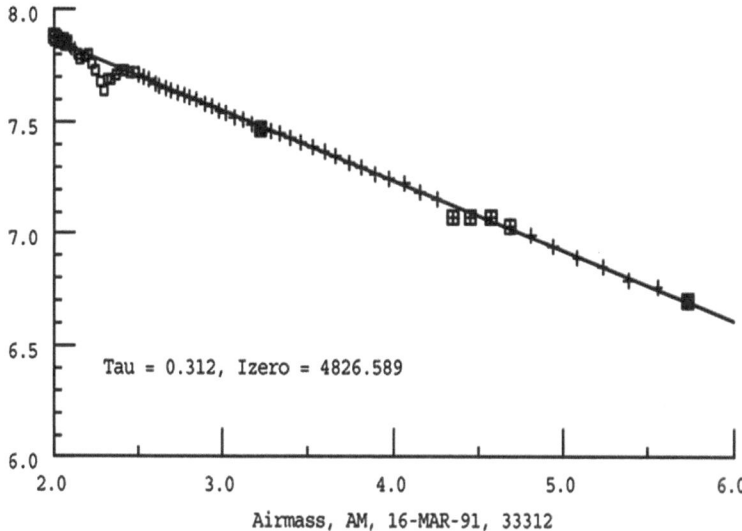

Fig. 2.7 Objective cloud-screening algorithm imposed in Langley calibration $\lambda = 500$ nm (Harrison and Michalsky 1994)

second iteration removes two points marked with a solid black box. Though these would not symmetrically affect the regression if retained, this second iteration is important if the data are noisier. At a difficult site, this method can provide a free long-term stability test for the instrument and permit the instrument calibration to be tied to solar output.

2.4.3.2   Clear-Sky Detection Algorithm

The concept of using statistical filter in Langley-plot as constrain for data selection is limited only for true clear-sky condition. In particular, under fictitious clear-sky conditions that are unable to be identified by the statistical filter, large inaccuracies may incur in the Langley extrapolation as part of the cloudy or aerosol-contained data will be used as guidance in the filtration process. Unlike the statistical-filtered Langley method that based on the simple minded notion of using least square regression on all available data, the clear-sky detection algorithm in Langley-plot uses only clear-sky condition data for the calibration analysis. In this way, the selection is completely automated and independent of the true clear-sky condition on a single day of observation.

   One of the most popular algorithms used in Langley calibration is the Long and Ackerman clear-sky detection algorithm (Augustine et al. 2003). The algorithm uses four sequential tests that scrutinize total solar and diffuse solar irradiance to detect cloud-free skies. These tests hypothesize that cloudy and hazy skies exhibit characteristics in the components of down-welling shortwave irradiance that clear

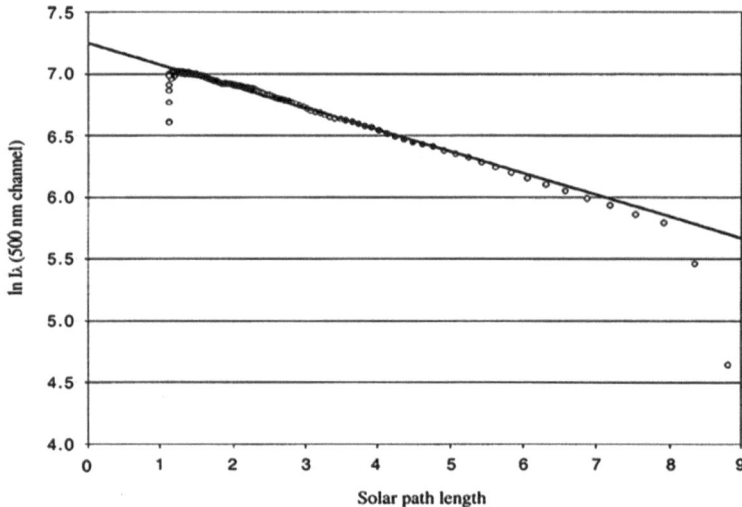

Fig. 2.8 Langley plot for the MFRSR 500 nm channel. *Solid circles* represent time periods identified as clear by Long and Ackerman clear-sky detection algorithm (Augustine et al. 2003)

skies do not. The first two tests eliminate periods of obvious cloudiness by comparing normalized transformation of the total and diffuse solar measurements to expected clear-sky limits. The other two tests examine temporal variations of parameters computed from the total and diffuse solar irradiance to further remove subtle periods of thin cloud or hazy conditions for more confident clear sky condition.

An example of a Langley plot for the MFRSR 500 nm channel at Table Mountain SURFRAD station the morning of 23 Apr 2001 is adapted from Augustine et al. (2003) given in Fig. 2.8. The solid circles represent time periods identified as clear by the Long and Ackerman (2000) method and the line is the least squares linear fit to the solid (clear sky) points only. The resultant $I_{o,\lambda}$ calibration value was applied to two Asian dust-related high air pollution events and the results suggested that error of retrieval is $\pm 0.01 \pm 0.05$, depending on the solar zenith angle. Though this method had been shown useful for selecting periods of MFRSR data appropriate for Langley-plot calibration, it requires collocated independent broadband solar component measurement for the clear-sky determination particularly for nominally un-calibrated radiometer.

2.4.4 Maximum Value Composite (MVC) Langley Method

Quiet recently, a new solar Langley calibration method to derive AOD from MFRSR data under extremely hazy atmospheric condition was proposed (Lee et al. 2010). It involves the acquisition of the maximum value composite (MVC) of

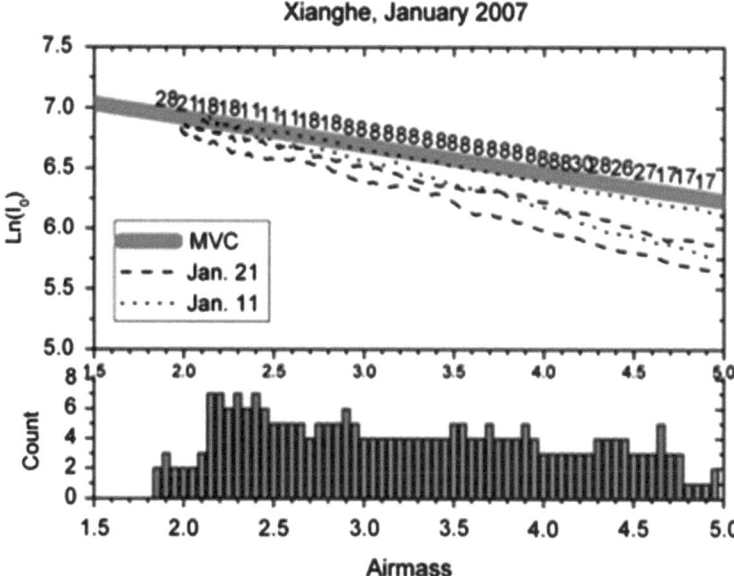

Fig. 2.9 Comparison between maximum value composite (MVC) Langley method and conventional Langley method at 500 nm channel (Lee et al. 2010)

the largest irradiance values at a given air mass. Regression of the Langley plot is based on these values because they can represent the clear-sky and minimum aerosol loading. Due to statistical uncertainties, not all maximum values can be used. Anomalous values such as zero or abnormally large/small values are removed by screening out local minima/maxima. This is performed by comparing values in neighboring air mass bins from which a standard deviation of relative difference of <1 % is removed.

Figure 2.9 shows the maximum value composite (MVC) result at 500 nm for one month of period and conventional Langley plots of single day within the month in dotted and dashed line (Lee et al. 2010). Histogram represents the number of days contributing to the MVC Langley method. All three resultant regression lines have high correlation coefficient >0.99 but different in y-intercept which represents the calibration value. Differences in I_0 by MVC regression and by single day Langley plots lead to large error (0.01 ∼ 0.40) in AOD determination. When compared to the AERONET method, results from MVC Langley method are comparable and within the acceptable error of <0.02 (Lee et al. 2010). However, one major shortcoming of this method is the MVC method cannot deal with temporal changes in extrapolated value of Langley-plot during a given composite period. To be specific, if the period of time is too long, information about the temporal variability is lost and if it is too short, there may be a dearth of valid data.

2.4.5 Comparative Langley Method

The comparative Langley method basically works in a recalibration basis by comparing to a well-calibrated wavelength or extrapolated value for improved calibration constant. Unlike the four previously described methods, this method does not perform the calibration experimentally instead it improves the calibration constant intrinsically through recalibration that based on presumed constant.

2.4.5.1 Ratio Estimation

It is a comparative calibration method which depends on a known calibration of a reference wavelength to permit calibration at the other wavelengths by assuming the relative size distribution of aerosol to remain constant as (Arai and Liang 2011)

$$\tau_a(\lambda, t) = \pi M(t) \int K_{ext}(r, \lambda) f(r') d \ln r', \qquad (2.45)$$

where $f(r')$ is the relative size distribution that is dependent only on the particle radius r', and $M(t)$ is the multiplier necessary to produce correct size distribution at some time, t. In this way, the ratio of aerosol optical depth between different wavelengths is assured to be constant as

$$\tau_a(\lambda_1, t)/\tau_a(\lambda_2, t) = \tau_a(\lambda_1, t_o)/\tau_a(\lambda_2, t_o) = \psi. \qquad (2.46)$$

Thus, calibration at other wavelengths is feasible using the reference wavelength that is assumed to be well-calibrated by

$$\ln P(\lambda_1) + m(\tau_m(\lambda_1) + \tau_o(\lambda_1)) = \ln P_o(\lambda_1) - \psi m \tau_a(\lambda_o), \qquad (2.47)$$

where λ_0, λ_1 are the reference and calibrate wavelength, respectively.

By adopting the similar approach suggested by Arai and Liang (2011), the improved Langley method by ratio estimation is conducted in three processes, level 0, calibration and level 1. In the level 0, based on the reference wavelength, the AOD at other wavelengths are estimated using the wavelength dependent relationship as Eq. 2.47. In the calibration, P_o at each observation is retrieved using the estimated AOD from level 0. Finally, in the level 1, the measured pixel is calibrated into direct normal irradiance (DNI) in physical unit using the P_o value obtained in the calibration level. Thus, more accurate solution of AOD can be estimated by reanalysis of the calibrated volume spectrum using the absolute extraterrestrial constant obtained directly from reference solar spectrum at top-of-atmosphere.

2.4.5.2 Monte Carlo Approximation

This comparative Langley method implements a weighted Monte Carlo (MC) approximation to find an improved calibration by minimizing the diurnal variation in Angstrom's exponent α and its curvature γ (Kreuter et al. 2013). The method simulates a large ensemble of random combination of calibration constant weighted with a Gaussian uncertainty function centered on a mean constant, and selects the calibration constant that yields the smallest diurnal variations (DV). The AOD dependency on wavelength is usually described by Angstrom's power law as

$$\log \tau_a = \log \beta_\mu - \alpha \log \lambda, \tag{2.48}$$

with λ is the wavelength in microns, β_μ is the AOD at wavelength of one micron. To account for a possible curvature, it has become common to add a quadratic term in logarithm λ as

$$\log \tau_a = \log \beta_\mu * + \alpha \log \lambda - \gamma \log^2 \lambda. \tag{2.49}$$

The parameters α and γ are determined by regression of Eqs. 2.48 and 2.49, respectively. The idea of this method is to harness the sensitivity of parameter α and γ on derived AOD by minimizing any residual DV in α and γ. Since both α and γ show independent DVs, the total diurnal variation (TDV) amplitude to be minimized as

$$TDV^2 = DV_\alpha^2 + DV_\gamma^2. \tag{2.50}$$

Noted that using a random number for an erroneous calibration constant V'_o that is normally distributed is an implicit weighting of the solution by the mean square error of the absolute V_o and MC calibration constant V_{mc}. Therefore, smaller deviations of V'_o from V_o are more likely generated in the MC approximation. Thus the retrieval of AOD using V_{mc} should be found as close to V_o as possible. The method had been proven to reduce the calibration uncertainty by a factor of up to 3.6 (Kreuter et al. 2013). It may also be easily generalized to other sunphotometer with more aerosol wavelength channels to improve the calibration beyond the Langley uncertainty.

References

Alexandrov MD, Marshak A, Cairns B et al (2004) Automated cloud screening algorithm for MFRSR data. Geophys Res Lett 31:L04118. doi:10.1029/2003GL019105

Arai K, Liang XM (2011) Comparative calibration method between two different wavelengths with aureole observations at relatively long wavelength. Int J Appl Sci 2:93–101

Augustine JA, Cornwall CR, Hodges GB et al (2003) An automated method of MFRSR calibration for aerosol optical depth analysis with application to an Asian dust outbreak over the United States. J Appl Meteorol 42:266–278

Bohren CF, Huffman DR (1983) Scattering by particles. Absorption and Scattering of Light by Small Particles

Che H, Zhang X, Chen H et al (2009) Instrument calibration and aerosol optical depth validation of the China Aerosol Remote Sensing Network. J Geophys Res 114:D03206. doi:10.1029/2008JD011030

Djamila H, Ming CC, Kumaresan S (2011) Estimation of exterior vertical daylight for the humid tropic of Kota Kinabalu city in East Malaysia. Renew Energy 36:9–15

Dubovik O, Holben B, Eck TF et al (2002) Variability of absorption and optical properties of key aerosol types observed in worldwide locations. J Atmos Sci 59:590–608

Dubovik O, King MD (2000) A flexible inversion algorithm for retrieval of aerosol optical properties from Sun and sky radiance measurements. J Geophys Res 105:20673–20696

Dubovik O, Smirnov A, Holben BN (2000) Accuracy assessments of aerosol optical properties retrieved from AERONET Sun and sky-radiance measurements. J Geophys Res 105:9791–9806

Harrison L, Michalsky J (1994) Objective algorithms for the retrieval of optical depths from ground-based measurements. Appl Opt 33:5126–32

King MD, Byrne DM, Herman BM, Reagan JA (1978) Aerosol size distributions obtained by inversion of spectral optical depth measurements. J Atmos Sci 35:2153–2167

Knobelspiesse KD, Pietras C, Fargion GS et al (2004) Maritime aerosol optical thickness measured by handheld sun photometers. Remote Sens Environ 93:87–106

Kreuter A., Wuttke S, Blumthaler M (2013) Improving Langley calibrations by reducing diurnal variations of aerosol Ångström parameters. Atmos Meas Tech 6:99–103. doi:10.5194/amt-6-99-2013

Lee KH, Li Z, Cribb MC et al (2010) Aerosol optical depth measurements in eastern China and a new calibration method. J Geophys Res 115:1–11. doi:10.1029/2009JD012812

Michalsky JJ, Kiedron PW (2008) Comparison of UV-RSS spectral measurements and TUV model runs for clear skies for the May 2003 ARM aerosol intensive observation period. Atmos Chem Phys 8:1813–1821. doi:10.5194/acp-8-1813-2008

Mie G (1908) Contributions to the optics of turbid media particularly of colloidal metal solutions. Ann Phys 330:377–445

Nieke J, Pflug B, Zimmermann G (1999) An aureole-corrected Langley-plot method developed for the calibration of HiRES grating spectrometers. J Atmos Solar-Terrestrial Phys 61:739–744

O'Neill NT, Eck TF, Holben BN, Smirnov A, Dubovik O, Royer A (2001) Bimodal size distribution influences on the variation of angstrom derivates in spectral and optical depth space. J Geophys Res Atmos 106:9787–9806

Schmid B, Wehrli C (1994) High precision of calibration of a sun photometer using Langley plots performed at Jungfraujoch (3580 m) and standard irradiance lamps. Geoscience and Remote Sensing Symposium 1994. IGARSS'94. Surface and Atmospheric Remote Sensing: Technologies, Data Analysis and Interpretation. International. pp 2314–2316

Shaw GE (1983) Sunphotometry. Bull Am Meteorol Soc 64:4–10

Slusser J, Gibson J, David B et al (2000) Langley method for calibrating UV filter radiometer. J Geophys Res 105:4841–49

Smirnov A, Holben BN, Eck TF et al (2000) Cloud-screening and quality control algorithms for the AERONET database. Remote Sens Environ 73:337–349

Smirnov A, Holben BN, Giles DM et al (2011) Maritime aerosol network as a component of AERONET–first results and comparison with global aerosol models and satellite retrievals. Atmos Meas Tech 4:583–597

Tanaka M, Nakajima T, Shiobara M (1986) Calibration of a sunphotometer of direct-solar and circumsolar by simultaneous radiations measurements. Appl Opt 25:1170–1176

Thomason LW, Herman BM, Reagan JA (1983) The effect of atmospheric attenuators with structured vertical distributions on air mass determinations and Langley plot analysis. J Atmos Sci 40:1851–1854

Chapter 3
Near-Sea-Level Langley Calibration Algorithm

Abstract As compared to other methods, measurement of aerosol optical depth (AOD) using sunphotometers offer several advantages. However, it suffers a drawback as calibration of the instrument required to be performed at high altitude due to temporal drifts in the atmospheric condition during the calibration. To solve this, a new Langley calibration algorithm has been designed for AOD measurement using spectroradiometer instrument. The key advantages of the proposed algorithm are its objectivity, computational efficiency and the ability to detect short intervals of cloud transits. It avoids travelling to high altitude mountain that the conventional calibration procedure always practiced for frequent calibration. Most importantly, neither it requires priori knowledge of the instrument calibration nor a collocated calibrated instrument for nominal calibration transfer to perform the cloud-screening procedure.

Keywords Near-sea-level calibration · Perez-Dumortier model · Clear-sky detection · Statistical filtration

3.1 Development of Near-Sea-Level Langley Calibration Algorithm

In the current practice, Langley calibration often needs to be determined from the data that have been ideally cloud-screened for accurate regression. Either for data collected from high altitude or near-sea-level sites, an effective cloud screening algorithm has an integral part in Langley calibration. Over the past decade, several algorithms had been developed for this purpose such as reject points that exhibit high derivatives on the statistical filter (Harrison and Michalsky 1994), or extract clear sky data by imposing thresholds on standard deviation of the measured values (Michalsky et al. 2001). However, the practice of using a least squares regression on all collected data works only under true clear-sky condition or at

J. Dayou et al., *Ground-Based Aerosol Optical Depth Measurement Using Sunphotometers*, SpringerBriefs in Applied Sciences and Technology, DOI: 10.1007/978-981-287-101-5_3, © The Author(s) 2014

least a large fraction of data represents clear-sky condition. To respond over this limitation, Long and Ackerman (2000) developed ensemble four sequential tests on the ratio between direct and diffuse broadband measurements to detect cloud-free skies based on the observation that cloudy skies exhibit characteristics in the components of downwelling shortwave irradiance that clear skies do not. However, to apply this algorithm on sunphotometers, a collocated broadband radiometer is necessary. Likewise, to apply on standalone MFRSR that measures diffuse and global irradiance in spectral bands and a broadband channel of the solar spectrum, its broadband channel must be calibrated for absolute irradiance though high accuracy is not necessary (Augustine et al. 2003). To perform this, temporarily running a calibrated broadband pyranometer alongside an MFRSR to transfer a nominal calibration to its broadband channel is all that is necessary for a standalone MFRSR. This would limit the practicality of the method since not all MFRSR and/or sunphotometer stations are also equipped with broadband pyranometer for this application. This means an effective cloud screening algorithm applicable in such a case should work on raw un-calibrated spectral measurement so that collocated broadband measurement is unnecessary. On this issue, another objective algorithm that based on spectral measurement was developed based on the degree of horizontal inhomogeneity of an atmospheric field, which does not require priori knowledge of the instrument calibration (Alexandrov et al. 2004).

On the ground of developing an effective calibration algorithm that works at near-sea-level and also on raw un-calibrated spectral measurement, the proposed calibration algorithm selects appropriate measurements suited for Langley calibration by imposing repetitive regression algorithm (RRA) on the un-calibrated spectral irradiance data until highest correlation in Langley plot is obtained. In this way, the contaminations of thick clouds and short-interval broken clouds are properly constrained without depending on the broadband measurement. In other words, the method introduces an objective algorithm to constrain the Langley extrapolation using the combination of clear-sky detection model and statistical filter. The former is to ascertain only cloudless and clear sky data is selected for the regression, and the latter is to further filter the resulting regression for improved instrument's response. Figure 3.1 shows the depiction of the proposed calibration algorithm. Detail of the proposed algorithm is discussed in the following.

3.1.1 Clear-Sky Detection Model

It is hypostasized that the clear sky conditions at high altitude can be accurately approximated at near-sea-level if there is a method to select such a data. For this purpose, Perez–Dumortier (PDM) model is used for clear-sky detection method for the selection of clear sky data. This model is selected because it had been proven to be appropriate in classifying the sky type and one of the most acknowledged precise model for predicting delighting and sky classification (Zain-Ahmed et al. 2002; Djamila et al. 2011).

Fig. 3.1 Illustration of the new Langley calibration algorithm using combination of clear-sky filtration model and statistical filter (Chang et al. 2014)

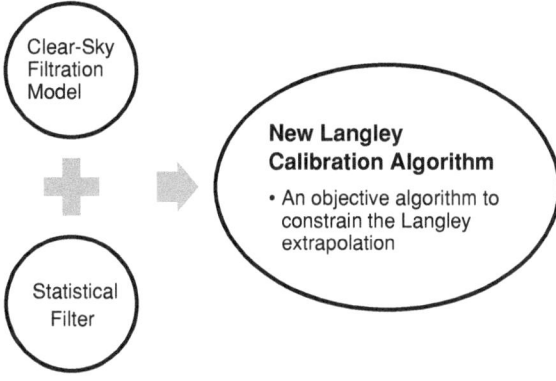

Table 3.1 Perez model classification of sky condition (Djamila et al. 2011)

Value of indices		Sky conditions
Sky ratio, *SR*	Clearness index, ε	
$SR \leq 0.30$	$\varepsilon \geq 4.50$	Clear sky
$0.30 < SR < 0.80$	$1.23 < \varepsilon < 4.50$	Partly cloudy or intermediate
$SR \geq 0.80$	$\varepsilon \leq 1.23$	Cloudy of overcast

In Perez model, the sky is classified into three type namely clear sky, partly cloudy or intermediate and cloudy of overcast (Table 3.1). It uses the Perez's clearness index ε as indicator of sky type which can be calculated using the relationship between the diffuse, I_{ed} and global, I_{eg} horizontal irradiance as governed by (Perez et al. 1990)

$$\varepsilon = \frac{((I_{ed} + I_{dir})/I_{ed}) + 1.041\varphi_H^3}{1 + 1.041\varphi_H^3} \tag{3.1}$$

where I_{dir} is the direct normal irradiance and ϕ_H is the solar zenith angle in radian.

In Dumortier model, the sky is classified into five types using Nebulosity index (*NI*) as indicator of sky type computed by the relationship between cloud ratio, diffuse irradiance I_{ed} and global irradiance I_{ed} which is (Zain-Ahmed et al. 2002):

$$NI = \frac{1 - I_{ed}/I_{eg}}{1 - CR}. \tag{3.2}$$

CR is the cloud ratio given as

$$CR = \frac{I_{d,cl}}{\left[I_{d,cl} + \exp(-4mAr)\sin\alpha\right]}, \tag{3.3}$$

Table 3.2 Dumortier model classification of sky condition (Zain-Ahmed et al. 2002)

Type of sky	Value of NI
Blue	$0.95 < NI < 1.00$
Intermediate blue	$0.70 < NI < 0.95$
Intermediate mean	$0.20 < NI < 0.70$
Intermediate overcast	$0.05 < NI < 0.20$
Overcast	$0.00 < NI < 0.05$

where $I_{d,cl}$ represents the clear sky illuminance is

$$I_{d,cl} = 0.0065 + (0.255 - 0.138 \sin \alpha_a) \sin \alpha_a \qquad (3.4)$$

and

$$Ar = \{5.4729 + m[3.0312 + m\{-0.6329 + m(0,091 - 0.00512m\}]\}^{-1}. \qquad (3.5)$$

Ar is the Rayleigh scattering coefficient, m is the optical air mass and α_a is the solar altitude. Table 3.2 shows the different type of sky condition according to nebulosity index.

The idea of combining these two models as an ensemble sky classification is based on the availability of both models for quantitative detection of very small changes in atmospheric turbidity. In most circumstances, cloud cover is the most dominant factor that determines the sky type. Under cloud-abundant condition, the sky type can be accurately classified based on the ratio of diffuse-to-global (D/G) irradiation that implicitly gives the amount of scattered light in the sky. This is due to the flux of sunlight is not wavelength dependent as the cloud droplets are larger than the light's wavelength and scatter all wavelengths approximately equally. Given that Dumortier model predicts the sky condition based on D/G ratio, it is expected to perform the best under cloudy and overcast sky where the scattering of sunlight is predominant. However, under intermediate and quite clear sky where diffuse irradiance is relatively lower, the D/G ratio becomes less dominant in sky type classification. In this case, Perez model is selected as a complementary model to classify the sky type. Unlike the former model, it uses the ratio of global-to-diffuse (G/D) irradiation to predict the sky condition that has less dependence on cloud-scattering effects. Besides, they are also computationally simple where only using the relationship between solar geometry, diffuse irradiance and global irradiance for determination of their values.

3.1.2 Statistical Filter

In conventional statistical filtration, the practice of rejecting points exhibit high derivatives (Harrison and Michalsky 1994) or imposing threshold on standard deviation of measured values (Michalsky et al. 2001) is only useful under true clear-sky condition. While the aim of the proposed algorithm is applicable at near-sea-level where cloud loadings are reasonably abundant, the simple notion of using least-squares regression on all data is irrelevant. In many cases at near-sea-level sites, the regression in Langley plot is not ideally singular even after removing the suspected cloudy periods. This occurs probably because of the contamination of varying AOD during the calibration period. An important goal of the proposed algorithm development is to arrive at a method that the selection is completely automated, and can be applied on a routine basis to the time-series data with no subjective preparation. On reaching that, the statistical filter is integrated on the PDM-filtered data for further screening to affect a more robust linear regression in Langley plot. More specifically, in the statistical filtration, two iterations are executed where the first is to perform a conventional least-square regression on the remaining points after PDM filtration, and the second is to compute the residuals of each points around the regression line then a sweep is performed to remove all points that have residual >0.006 around the regression line (Harrison and Michalsky 1994). The idea behind this statistical filter is to eliminate possible outliers and instabilities due to the instrument responses for very small derivatives that are unable to be detected by the PDM algorithm. In addition, this error estimator is a ratio of intensities, and hence it is independent of both the evolution of air mass as well as the absolute calibration of the detector.

3.2 Implementation of the Proposed Calibration Algorithm

Basically, the implementation of the proposed calibration algorithm is a two-stage screening process where the first stage is for clear-sky PDM filtration and the second is statistical filtration. More specifically, Fig. 3.2 illustrates the full depiction of the implementation of PDM filtration. Then, the corresponding clearness index (ε) and nebulosity index (NI) for each data is computed using Eqs. 3.1 and 3.2. After that, multiple permutated criteria of ε and NI are generated in a given resolution step of 0.01 for each index. These indexes are then used as the criteria for optimum clear-sky filtration in a repetitive regression algorithm (RRA). The filtration occurs iteratively until the highest correlation R^2 in Langley plot is obtained for which determines the threshold value ε and NI of clear-sky condition. After the clear-sky filtration, a statistical filtration step as described in Sect. 3.1.2, is imposed to the resulting Langley plot to yield the final filtration product. Finally, a regression is feasible to determine the extraterrestrial constant at zero air mass using the Langley extrapolation technique. This step is generally to eliminate other measurement errors and uncertainties.

On the whole, the PDM calibration algorithm constrains the Langley extrapolation based on repetitive regression algorithm using multiple permutated criteria (NI_x, ε_x) until the best linearity between natural logarithm of light intensity and air mass is obtained. The advantage of using only clear-sky data defined by the proposed algorithm is that noise is reduced and a confident extrapolation to zero

Fig. 3.2 Flowchart of the implementation of PDM filtration in repetitive regression algorithm (RRA)

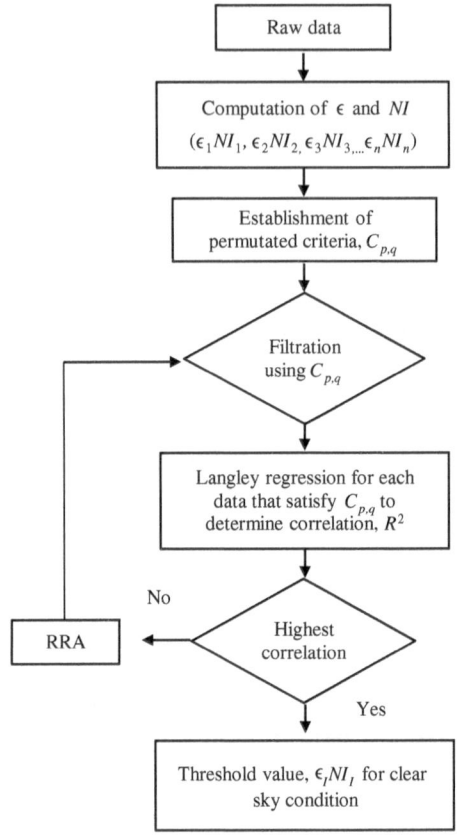

air mass by simple regression is feasible. Instead of using a single day for Langley plot, the method collects several days of clear-sky Langley plots over a period of time. In this way, it produces a pool of extrapolated values so that a reliable mean can be computed. Since the proposed algorithm depends on the highest correlation plotted from the pool of screened data, it also increases the confidence that the mean extrapolated value is stable and free of any effects that promote changes in thin cirrus cloud and aerosol loading.

References

Alexandrov MD, Marshak A, Cairns B et al (2004) Automated cloud screening algorithm for MFRSR data. Geophys Res Lett 31:L04118. doi:10.1029/2003GL019105
Augustine JA, Cornwall CR, Hodges GB et al (2003) An automated method of MFRSR calibration for aerosol optical depth analysis with application to an Asian dust outbreak over the United States. J Appl Meteorol 42:266–278

Chang JHW, Dayou J, Sentian J (2014) Development of near-sea-level calibration algorithm for aerosol optical depth measurement using ground-based spectrometer. Aerosol Air Qual Res 14:386–395

Djamila H, Ming CC, Kumaresan S (2011) Estimation of exterior vertical daylight for the humid tropic of Kota Kinabalu city in East Malaysia. Renew Energy 36:9–15

Harrison L, Michalsky J (1994) Objective algorithms for the retrieval of optical depths from ground-based measurements. Appl Opt 33:5126–5132

Long CN, Ackerman TP (2000) Identification of clear skies from broadband pyranometer measurements and calculation of downwelling shortwave cloud effects. J Geophys Res 105:609–626

Michalsky JJ, Schlemmer JA, Berkheiser WE et al (2001) Multiyear measurements of aerosol optical depth in the Atmospheric Radiation Measurement and Quantitative Links programs. J Geophys Res 106:12009–12107

Perez R, Ineichen P, Seals R et al (1990) Modelling daylight availability and irradiance components from direct and global irradiance. Sol Energy 44:271–289

Zain-Ahmed A, Sopian K, Abidin ZZ, Othman MYH (2002) The availability of daylight from tropical skies—a case study of Malaysia. Renew Energy 25:21–30

Chapter 4
Implementation of Perez-Dumortier Calibration Algorithm

Abstract To avoid the unnecessary needs to travel to high altitude for sunphotometers calibration, Perez-Dumortier calibration algorithm has been used as an objective means to select the right intensity data so that the calibration can be performed at any altitude levels. The governing theory of the algorithm was discussed in the previous chapter. This chapter presents information on how to implement the Perez-Dumortier calibration algorithm using actual field measurement. The implementation of the filtration procedure in step-by-step is discussed to render better framework of the proposed calibration algorithm. The aerosol retrieval inversion uses the extraterrestrial constant obtained from the final Langley plot to calculate retrieved AOD. The implementation example uses irradiance-matched technique by i-SMARTS radiative transfer code to derive corresponding reference AOD for validation purposes. The reliability of the technique was substantiated by radiative closure experiment to verify the promising direct solar irradiance to accurately derive the reference AOD values.

Keyword Langley extrapolation · Irradiance-matched · SMARTS · Radiative transfer model

4.1 Instrumentation

Measurement of AOD can be performed using ground-based sunphotometers (e.g. CIMEL, MFRSR) and spectral/broadband radiaometers (e.g. Licor spectroradiometer, pryanometer). These instruments typically measure spectral or broadband solar radiation ranges from near-UV to near-IR wavelengths. As an example of these instruments, ASEQ LR-1 spectrometer is a portable, compact and robust spectral radiometer that measures light intensity over a specific portion of electromagnetic spectrum. The variable measured is the light's intensity at a given wavelength that is invariant. Table 4.1 shows the specifications of the spectrometer. This instrument has a 3648-element CCD-array silicon photodiode detector

J. Dayou et al., *Ground-Based Aerosol Optical Depth Measurement* 39
Using Sunphotometers, SpringerBriefs in Applied Sciences and Technology,
DOI: 10.1007/978-981-287-101-5_4, © The Author(s) 2014

Table 4.1 Specifications of ASEQ LR-1 Spectrometer

Specifications	ASEQ LR-1 Spectrometer
Spectral range	300–1100 nm
Spectral resolution (FWHM)	$< \sim 1$ nm (with 50 μm slit)
Weight	430 grams
Dimension	102 mm x 84 mm x 59 mm
Detector	Toshiba TCD1304AP linear silicon CCD array
A/D resolution	14 bit
Fiber optic connector	SMA 905 to 0.22 numerical aperture single strand optical fiber
CCD reading time	14 ms
Cosine corrector	$0.5°$

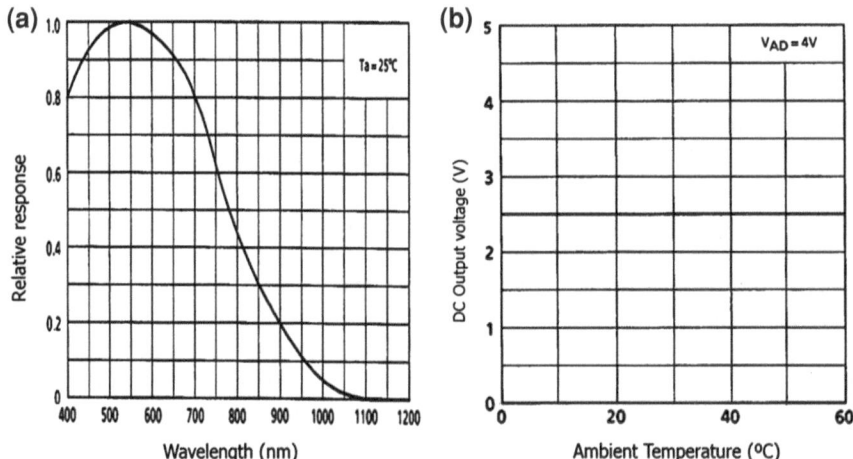

Fig. 4.1 Typical performance curves of Toshiba TCD1304 linear silicon CCD array in **a** spectral response and **b** ambient temperature as provided in datasheet

from Toshiba that enables optical resolution as precise as 1 nm (FWHM). Figure 4.1a presents the photodiode sensitivity response within the measured spectral range. It has relative response peaks in the mid-visible range and gradually decreases toward the near-infrared wavelengths. The output of the sensor is not prone to degradation for temperatures up to 60 °C as provided in the sensor's datasheet (Fig. 4.1b).

To facilitate a better understanding on the PDM calibration algorithm, an actual field measurement data that conducted at Tun Mustapha Tower, Kota Kinabalu, Sabah, Malaysia (116 °E, 6 °N) is used in the step-by-step demonstration on the implementation. This site was selected due to its location, which is near-sea-level with site altitude of 7.8 m, permitting not only the investigation of the feasibility of Langley calibration at low altitude but also the observed spectrum is obstruct-free from irrelevant objects such as trees or artificial buildings. In Langley-plot calibration method, measurements are typically made within the air mass range from 2 to 6 for the visible band wavelengths. This range of air mass is associated to high zenith angle SZA 60.0° to 80.4° which corresponds to the local time from 0640 to 0815 h over the selected study area. This range of air mass is used because it provides sufficient points of observation for more accurate regression in Langley plot with less extrapolation error. On the other hand, higher air masses are avoided due to greater uncertainty in air mass caused by refraction that is increasingly sensitive to atmospheric temperature profiles (Chang et al. 2013).

4.2 Determination of Langley Extraterrestrial Constant Using the Proposed Calibration Algorithm

In this example, a 2-month period of data measurements from April to May of 2012 was used. Within this period, a total of 730 data had been collected. However, not all data can be used for Langley regression as the part of measurements may be contaminated by cloud cover, aerosol loading, and sun-pointing errors. Therefore, the clear-sky detection PDM model is used to identify these points and filter them.

For each data (hereinafter denoted as D_n, where n represents the number of observations), the corresponding clearness index (ε) and nebulosity index (NI) can be computed using Eqs. 4.1 and 4.2, respectively and they are listed as follows

$$
\begin{bmatrix} D_1 \\ D_2 \\ \vdots \\ D_n \end{bmatrix} = \begin{bmatrix} \varepsilon_1 & NI_1 \\ \varepsilon_2 & NI_2 \\ \vdots & \vdots \\ \varepsilon_n & NI_n \end{bmatrix} \tag{4.1}
$$

The raw data of clearness and nebulosity index in Eq. 4.1 is here-forth known as clearness-nebulosity index (CNI). The CNI values are inserted into a repetitive regression algorithm for data filtration using permutated criteria, $C_{p,q}$ in the range of $1.23 \leq p \leq 1.89$ and $0.70 \leq q \leq 0.99$, where p and q represent the criteria index. The permutated criteria $C_{p,q}$ is repeated for other value of ε and NI at a step of 0.01. As a result, a series of permutated criteria is generated as

$$
C_{p,q} = \begin{bmatrix} C_{1,1} & C_{2,1} & \cdots & C_{67,1} \\ C_{1,2} & \ddots & & \vdots \\ \vdots & & \ddots & \vdots \\ C_{1,30} & \cdots & \cdots & C_{67,30} \end{bmatrix}. \tag{4.2}
$$

The criteria in Eq. 4.2 have their corresponding conditional value given in Eq. 4.3.

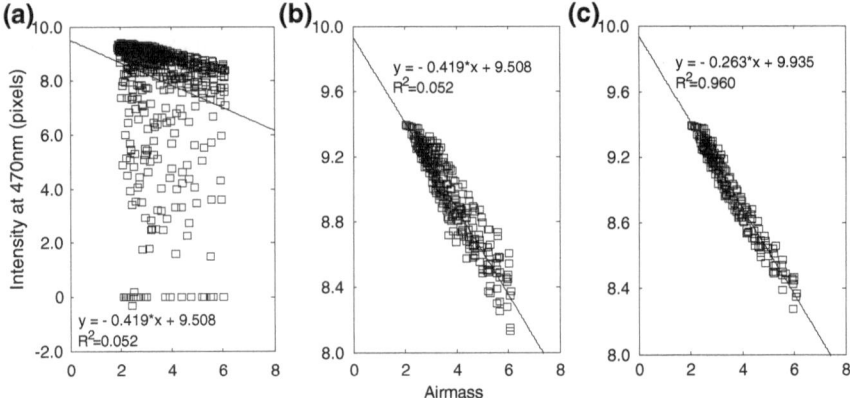

Fig. 4.2 Langley plot at 470 nm **a** before filtration, after filtration using (**b**) Perez-Dumortier model (NI $\geq$ 0.92 and $\epsilon \geq$ 1.55), and **c** statistical filtration

$$
[\varepsilon_p NI_q] = \begin{bmatrix} \varepsilon_1 \geq 1.23, NI_1 \geq 0.70 & \varepsilon_2 \geq 1.24, NI_1 \geq 0.70 & \cdots & \varepsilon_{67} \geq 1.89, NI_1 \geq 0.70 \\ \varepsilon_1 \geq 1.23, NI_2 \geq 0.71 & \ddots & & \vdots \\ \vdots & & \ddots & \vdots \\ \varepsilon_1 \geq 1.23, NI_{30} \geq 0.99 & \cdots & \cdots & \varepsilon_{67} \geq 1.89, NI_{30} \geq 0.99 \end{bmatrix}.
$$

$$(4.3)$$

As an example for implementation purpose, when criterion $C_{1,1}$ is imposed in the algorithm, only data D_n with value $\varepsilon_1 \geq 1.23$ and $NI_1 \geq 0.70$ will be used for the regression of Langley plot. As a result, some of the corresponding intensity data of D_n (intensity of the D_n data that does not fulfill the corresponding condition in Eq. 4.3) will be filtered out and the remaining are plotted against air mass to get the corresponding Langley plot. The corresponding correlation value of the Langley regression plot is then obtained. This step is repeated for each criterion given in Eqs. 4.2 and 4.3 until the highest correlation in the Langley plot is obtained. This is then followed by the statistical filtration as described in Sect. 3.1.2. It is imposed to the resulting Langley plot to eliminate other measurement errors and uncertainties.

To illustrate the resulting Langley plot of the proposed calibration algorithm, an example is shown in Fig. 4.2 for wavelength at 470 nm. Figure 4.2a shows the Langley plot of unfiltered data that consists of 730 points. After the repetitive regression algorithm is implemented, the data point reduced to 272 and the corresponding Langley plot is shown in Fig. 4.2b. Finally, the number of data further reduced to 200 after the statistical filtration is implemented, and the corresponding Langley plot is shown in Fig. 4.2c. The final Langley plot of the completely filtered data at given wavelength is then can be used to determine the extraterrestrial value and calibration factor.

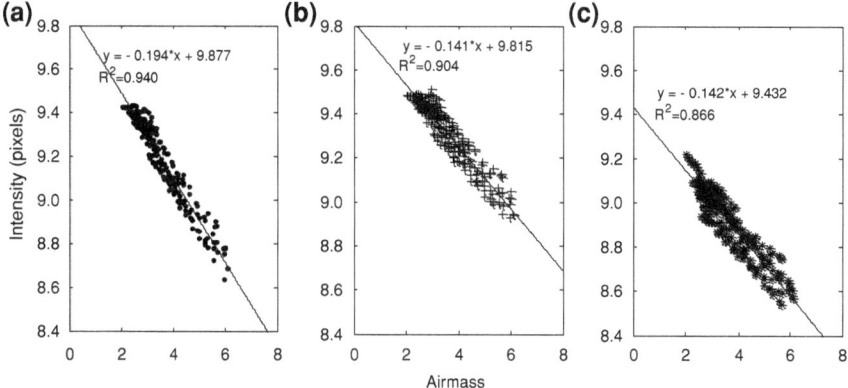

Fig. 4.3 Langley plot at (**a**) 500 nm, (**b**) 550 nm, and (**c**) 660 nm after Perez-Dumortier model and statistical filtration

Table 4.2 Resulting Langley plots after PDM and statistical filtration, n represents remaining data points where total initial point is 730, R^2 is correlation coefficient

Wavelength	Filtration	Number of data, n	Regression line	R^2
470 nm	$NI \geq 0.92$ and $\varepsilon \geq 1.55$	272	y = −0.262x + 9.941	0.883
	$\sigma < \pm 2\sigma$	200	y = −0.263x + 9.935	0.960
500 nm	$NI \geq 0.92$ and $\varepsilon \geq 1.55$	272	y = −0.194x + 9.887	0.840
	$\sigma < \pm 2\sigma$	214	y = −0.194x + 9.877	0.940
550 nm	$NI \geq 0.92$ and $\varepsilon \geq 1.55$	272	y = −0.138x + 9.812	0.768
	$\sigma < \pm 2\sigma$	221	y = −0.141x + 9.815	0.904
660 nm	$NI \geq 0.92$ and $\varepsilon \geq 1.55$	272	y = −0.131x + 9.386	0.713
	$\sigma < \pm 2\sigma$	221	y = −0.142x + 9.432	0.866

Figure 4.3 shows the final Langley plot for other wavelengths at 500 nm, 550 nm, and 660 nm. The detailed information of the Langley plot for each wavelength is shown in Table 4.2 for clarity. One can also be assured of the adequacy of data collection from this table. It is obvious that the final filtration product yields almost two-third of the data for all wavelengths which strictly complies with the fixed error criteria where a minimum of 1/3 of the pre-filtered data points should remain after the statistical filtration.

Table 4.3 summarizes the conclusive results of the Langley extrapolation to zero airmass for each studied wavelength. The calibration factor k for each wavelength is computed by dividing the corresponding extrapolated value with extraterrestrial constant at TOA obtained from ASTM G173-03 Reference Spectra. Multiplication of pixels measured at ground with this factor converts the measurements into physical unit in $W/m^2/nm$.

Table 4.3 Determination of calibration factor, k using ASTM G173-03 Reference Spectra

Wavelength (nm)	Extrapolated value P_o	Extraterrestrial constant (W/m^2/nm)	Calibration factor, k
470	9.935	1.939	9.400E-05
500	9.877	1.916	9.840E-05
550	9.815	1.863	1.020E-04
660	9.432	1.558	1.250E-04

4.3 Retrieval of Spectral AOD

In work example presented above, the study of spectral AOD is limited to visible band 400–700 nm (470, 500, 550 and 660 nm), where the only components that show non-negligible absorption are Rayleigh, ozone and nitrogen dioxide (Utrillas et al. 2000). In this way, contributions by all other constituents in the parameterization of total optical depth are negligible and hence errors can be reduced to minimum. Therefore, the total optical depth, $\tau_{\lambda,i}$ is hence governed only by Rayleigh, ozone and aerosol, which can be written as

$$\tau_{\lambda,i} = \tau_{R,i} + \tau_{o,i} + \tau_{a,j} \tag{4.4}$$

The Rayleigh contribution can be approximated using the relationship (Knobelspiesse et al. 2004)

$$\tau_{R,\lambda,i} = k_{Ray}(\lambda)\frac{p}{p_o}\exp\left(-\frac{H}{7998.9}\right) \tag{4.5}$$

where $k_{Ray(\lambda)}$ is the Rayleigh scattering coefficient, p is the site's atmospheric pressure, p_o is the mean atmospheric pressure at sea-level and H is the altitude from sea-level in meter. Similarly, ozone optical depth can be calculated using satellite observation of ozone in Dobson unit (DU) which is computed by (Knobelspiesse et al. 2004):-

$$\tau_{o,\lambda,i} = Zk_{oz}(\lambda)\times2.69e16\,\text{molecules}\big/\text{cm}^2 \tag{4.6}$$

where Z is ozone concentration in DU, $k_{oz(\lambda)}$ is ozone absorption cross section. Using the inverse technique, AOD is hence retrievable from $\tau_{\lambda,I}$ after eliminating the effects of other relevant atmospheric constituents, which in this case are Rayleigh and ozone contribution.

By using the extraterrestrial constant, P_o derived from the proposed algorithm in Table 4.3, the values of AOD at all wavelengths were retrieved for each observation using Eqs. 4.4–4.6. Unlike the Langley calibration analysis that requires homogenous stable atmospheric condition, retrieval of AOD is feasible as long as the observed sky is cloudless, which in this case produces a large pool of

Table 4.4 Database subset of retrieved optical depths using the proposed calibration algorithm over study area at Tun Mustapha Tower, Kota Kinabalu, Sabah, Malaysia in Apr–May 2012 (Total data n = 568)

Time	Retrieved optical depths											
	470			**500**			**550**			**660**		
	τ_R	τ_o	τ_a	τ_R	τ_o	τ_a	τ_R	τ_o	τ_a	τ_R	τ_o	τ_a
7-Apr-2012												
0652	0.184	0.003	0.077	0.143	0.008	0.051	0.097	0.023	0.038	0.046	0.015	0.077
0655	0.184	0.003	0.073	0.143	0.008	0.046	0.097	0.023	0.035	0.046	0.015	0.077
0658	0.184	0.003	0.083	0.143	0.008	0.054	0.097	0.023	0.041	0.046	0.015	0.085
0701	0.184	0.003	0.084	0.143	0.008	0.052	0.097	0.023	0.040	0.046	0.015	0.085
0707	0.184	0.003	0.084	0.143	0.008	0.051	0.097	0.023	0.038	0.046	0.015	0.089
0710	0.184	0.003	0.085	0.143	0.008	0.051	0.097	0.023	0.038	0.046	0.015	0.091
0713	0.184	0.003	0.082	0.143	0.008	0.047	0.097	0.023	0.034	0.046	0.015	0.089
0716	0.184	0.003	0.080	0.143	0.008	0.045	0.097	0.023	0.033	0.046	0.015	0.089
0719	0.184	0.003	0.082	0.143	0.008	0.045	0.097	0.023	0.032	0.046	0.015	0.091
0725	0.184	0.003	0.076	0.143	0.008	0.040	0.097	0.023	0.029	0.046	0.015	0.087
0728	0.184	0.003	0.081	0.143	0.008	0.044	0.097	0.023	0.032	0.046	0.015	0.091
0731	0.184	0.003	0.083	0.143	0.008	0.045	0.097	0.023	0.033	0.046	0.015	0.094
0734	0.184	0.003	0.085	0.143	0.008	0.048	0.097	0.023	0.037	0.046	0.015	0.096
0737	0.184	0.003	0.083	0.143	0.008	0.048	0.097	0.023	0.036	0.046	0.015	0.094
0740	0.184	0.003	0.097	0.143	0.008	0.059	0.097	0.023	0.047	0.046	0.015	0.113
0746	0.184	0.003	0.099	0.143	0.008	0.065	0.097	0.023	0.051	0.046	0.015	0.114
0749	0.184	0.003	0.103	0.143	0.008	0.068	0.097	0.023	0.055	0.046	0.015	0.116
0755	0.184	0.003	0.098	0.143	0.008	0.071	0.097	0.023	0.057	0.046	0.015	0.104
...	...	...	...	...	...	...	...	...	...	...	...	...

τ_R: Rayleigh optical depth, τ_o: ozone optical depth, τ_a: aerosol optical depth

useful data $n = 568$. Due to the large amount of data available, only selected data was presented in Table 4.4. These values are hereinafter termed as retrieved AOD for easy discussion.

4.4 Validation of the Proposed Calibration Algorithm

4.4.1 Irradiance-Matched by i-SMARTS Radiative Transfer Code

In order to validate the AOD retrieved from the proposed calibration algorithm, inverse-radiative transfer model (RTM) i-SMARTS can be used to simulate multiple AOD values, at distinct air mass and direct spectral irradiance (DSI). The comparison between the calculated AOD from radiative transfer model and the retrieved AOD from the proposed algorithm allows investigating the performance of the proposed calibration algorithm. It basically works in a look-up-table (LUT) approach by matching the measured irradiance with the calculated irradiance at given air mass with assumption that conditions of the atmosphere do not change significantly from the input parameters inserted into the RTM.

The i-SMARTS model is originated from SMARTS model which is a simplified RTM used to predict solar spectrum radiation under clear-sky condition when certain meteorological parameters are known. The use of SMARTS model is not limited to solar beam prediction but also useful in AOD retrieval when it is inversed from irradiance measurement. Its working principle is based on repeating RTM runs until convergence with measured data is achieved (Seidel et al. 2012). Similar to SMARTS simulation, this inverse model requires a priori knowledge of aerosol particle size distribution and spectral aerosol refractive indices. In our case, Shettle and Fenn Urban (SFU) aerosol model is used in the simulation because it exhibits significantly lower deviation independent from the input parameters of either aerosol optical depth or Ångström turbidity coefficient (Kaskaoutis and Kambezidis 2008). On selecting the aerosol model that best represents the aerosol type over the study area, the composition of atmospheric aerosol is used as reference.

In Shettle & Fenn model, rural aerosol model is intended to represent the aerosol under conditions where it is not directly influenced by urban and/or industrial aerosol sources. The rural aerosols are assumed to be composed of a mixture of 70 % water soluble substance (ammonium and calcium sulfate and also organic compounds) and 30 % dust-like aerosols (Shettle and Fenn 1979). In urban areas, air with a rural aerosol background is primarily modified by the addition of aerosols from combustion products and industrial source. Therefore, urban aerosol model is taken to be a mixture the rural aerosol (80 %) with carbonaceous soot-like aerosols (20 %). For maritime aerosol model, the aerosol compositions and distributions of oceanic origin are significantly different from continental aerosol types. These aerosols are largely sea-salt particles which are produced by the evaporation of sea-spray droplets and then have grown again due to aggregation of water under high relative humidity conditions.

In a study by Trivitayanurak et al. (2012), the composition and variability of atmospheric aerosol over Borneo had been studied using GEOS-Chem Global 3-D chemistry model in conjunction with aircraft and satellite observation. The result findings revealed that Borneo was a net exporter of primary organic aerosol and black carbon aerosol. The time-series of MODIS and model AOD over Borneo also suggests most of the regional aerosol attributes to sulphur, organic carbon, sea salt and a small source from black carbon and dust-like aerosol. Another local study by Sumari et al. (2009) reported that the characteristics of aerosols over the study area were highly influenced by anthropogenic species resulted from synthetic fertilizers and automobile exhaust. Given that the rural and maritime aerosol model both represent urbanized-free sources and largely sea-salt particles respectively, thus it is unlikely to represent the aerosol type over the study area using these models. Likely in the presence of carbonaceous aerosols and

Table 4.5 Input parameters for the AOD retrieval using i-SMARTS

Input parameters	Description
Altitude	7.844 m
Latitude	6.01 °N, 116.13 °E
Height above ground	0.934 m
Reference atmosphere	Tropical (summer/spring season)
Relative humidity	75 %
Instantaneous temperature	299.5 K
Regional albedo	0.31
Aerosol model	Shuttle & Fenn Urban (SFU) model
Solar constant	1366.1 $W/m^2/nm$
CO_2 mixing ratio	394.01 ppmv—provided by NOAA/ESRL
Ozone concentration	0.2611 atm-cm—provided by NASA, Giovanni

anthropogenic inputs, it is apparent that urban aerosol model is the best selection amongst them.

Other required input parameters are the atmospheric pressure, air temperature, relative humidity, zenith angle, azimuth, etc. Table 4.5 presents the input parameters inserted into i-SMARTS model for the AOD retrieval. In Table 4.5, relative humidity and temperature were default values set in the tropical reference atmosphere in the model. The use of default values in our case is because the hourly RH and temperature data provided by local meteorological department were sometimes missing due to unforeseen circumstances. Thus, it creates serious gap for averaging these variables. Therefore, these two variables are assumed constant as the averaged value in the reference tropical atmosphere. In this way, it reduces not only uncertainties in averaged value for existing gaps but also secures fast and effective retrieval to save computational time and extensive look-ups. Other parameters such as concentration of CO_2, O_3 and regional albedo that cannot be obtained from local meteorological were averaged according to the spatial and temporal valuation of the study area using satellite observations provided by NOAA-ESRL Physical Sciences Division, Boulder Colorado (URL: http://www. esrl.noaa.gov/psd/), NASA/Giovanni (Acker and Leptoukh 2007) and NASA/ Goddard Space Flight Center (URL: http://svs.gsfc.nasa.gov/goto?30371).

The AOD retrieval with i-SMARTS uses a look-up table (LUT) approach to allow fast convergence with the measured irradiance to calculate the AOD to be retrieved. The restrained LUT consists of two main parameters that determine the direct spectral irradiance (DSI), which are air mass and AOD (Fig. 4.4). The air mass interval between two successions is sampled at step of 0.1 from 2.0 to 6.0, corresponding to the observation period of the study. Meanwhile the AOD interval is simulated at a sampling rate 0.01 between 0.00 and 0.40. Based on our preliminary results, the average value of retrieved AOD is 0.108 with std. dev. 0.062 (min 0.008, max 0.322). Thus, an upper limit of 0.40 could at least guarantee an exceptional error of 20 % in the AOD derivation using the LUT. In addition, other study in Trivitayanurak et al. (2012) also reported the similar observation that

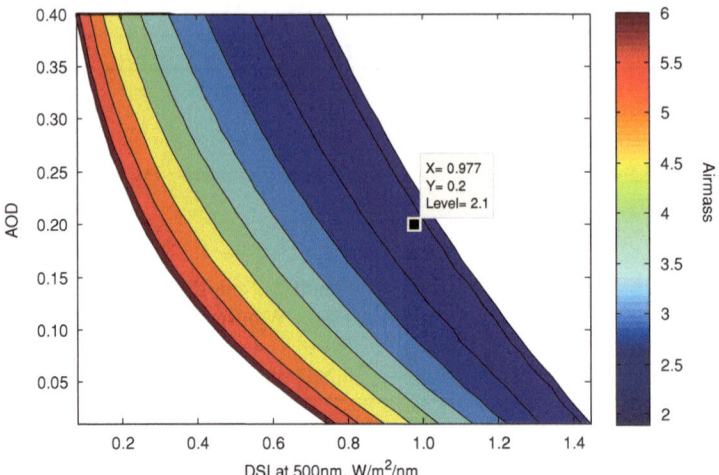

Fig. 4.4 Synthetic data of DSI at 500 nm from i-SMARTS in contour plot. The color bar represents air mass evolution from 2 to 6. The interpolation technique matches the measured DSI with the calculated DSI (*X-axis*) at distinct air mass (*Level*) to derive reference AOD (*Y-axis*)

AOD over Borneo is typically <0.4, peaking in February due to a transport pattern from the China region that affects the Southeast Asian region. The current LUT used in this method allows variation in AOD in the range of 0–0.4 and air mass in 2–6 for multiple synthetic DSI values. In ideal circumstance, other variables such as RH, O_3 and temperature should also be allowed to vary diurnally and on daily basis, but still this change could be very small and insignificant particularly in our example case that extends only in approx. 2 h (0630–0830) in 2 months of time. Moreover, when these variables (e.g. RH, O_3 etc.) are taken into account in the LUT, the existing database would be fold-up by the total sampling number of these variables. Definitely, this would result to heavy computational time and extensive look-ups. Although it is possible to integrate more dimensions in the LUT for long-term consideration, in the preliminary investigation stage, it is believed that the LUT should focus on the most two significant variables (AOD and air mass) for fast and effective retrieval.

The initiation of search requires matching air mass as the preliminary criteria. Then, convergence of the measured DSI with the calculated DSI values is done using the interpolation technique. This technique allows fast and accurate determination of AOD to be retrieved based on the exponential regression obtained from the inter-plot between DSI and AOD at given air mass. Using this LUT database, AOD is hence retrievable by matching the measured DSI with the simulated values at distinct air mass for each observation to allow point-by-point validation.

4.4.2 Radiative Closure Experiment

The basis of the LUT validation approach described lies on the promising accuracy of well calibrated irradiance to accurately calculate the truth-value of reference AOD. Given that our previous work in Chang et al. (2014) had validated the accuracy of measured irradiance in a radiative closure experiment (RCE), therefore the calculated AODs from this irradiance-matched approach should exhibit an acceptable accuracy of ~ 3 % on average. Figure 4.5 supported this by showing the measured irradiance matches the simulated value obtained in RCE with high correlation R2 > 0.8 and small error NMSE < 3 % for all wavelengths. Accordingly, the closure between measured and calculated irradiance at a given air mass derives the reference AOD value that best represents the actual turbidity condition, which in this case provides a reliable training data for the validation purposes. In other words, the reference AOD here represents the output of the irradiance-matching between measured and calculated irradiance in the LUT at distinct air mass. They are used for the validation purposes to validate the AOD retrieved from the proposed algorithm. In addition, the application of this irradiance-matched validation is also extended to other wavelengths. By matching the measured DSI at a single wavelength, I_λ in the radiative transfer calculation, only one single $\tau_{a(\lambda)}$ at that particular wavelength is calculated. Then, recalculation of RTM based on the spectral aerosol coefficients is executed to re-calculate multiple τ_a at other wavelengths, $\lambda_{i,j,k...n}$ for the validation purposes as

$$I_\lambda\left(m, \tau_{a,\lambda}\right) \overset{matched}{\leftrightarrow} RTM\left(I_{o,\lambda} e^{-(m,\tau_{a,\lambda})}\right) \overset{retrieve}{\longrightarrow} \tau_a(\lambda) \overset{RTM}{\longrightarrow} \begin{bmatrix} \tau_a(\lambda_i) \\ \tau_a(\lambda_j) \\ \tau_a(\lambda_k) \\ \vdots \\ \tau_a(\lambda_n) \end{bmatrix}$$

where m represents the distinct air mass at a given observation. In this way, the interrelated retrieval of AOD at other wavelengths, $\lambda_{i,j,k}$ is constrained and thus offers validation at multiple wavelengths, λ_n.

4.4.3 Performance Analysis

Using the interpolation technique described above, multiple AOD values are derived for each observation and these data are denoted as reference AOD to avoid confusion and tabulated in Table 4.6. It is clear that significant discrepancies are observed between retrieved and reference AOD particularly at 660 nm. The details of these discrepancies are presented in Fig. 4.6. Significant deviation from the reference value with poor correlation $R^2 > 0.79$ and relatively higher RMSE 0.071 is observed at wavelength 660 nm, and slight deviation is also observed at 470 nm

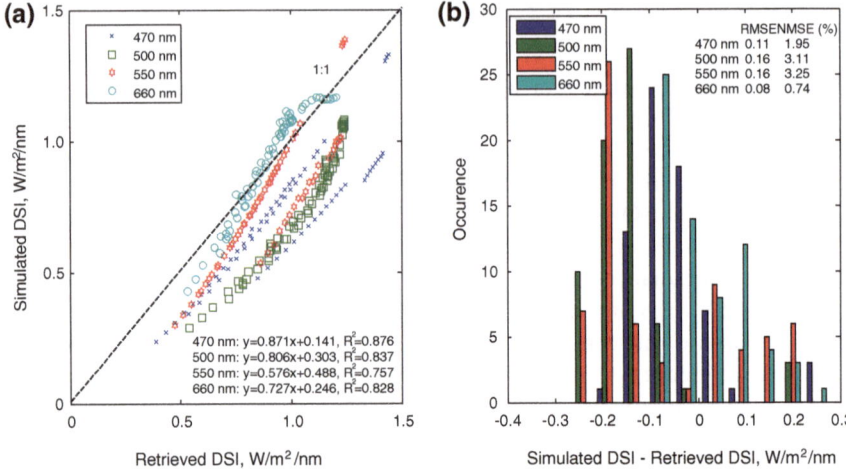

Fig. 4.5 Comparison between retrieved and simulated DSI in radiative closure experiment in **a** scatter plot and **b** histogram plot. (RMSE: root mean square error, NMSE: normalized mean square error)

Table 4.6 Database subset of reference optical depths simulated using i-SMARTS model over study area at Tun Mustapha Tower, Kota Kinabalu, Sabah, Malaysia in Apr–May 2012 (Total data = 568)

Time	Reference optical depths											
	470			500			550			660		
	τ_R	τ_o	τ_a	τ_R	τ_o	τ_a	τ_R	τ_o	τ_a	τ_R	τ_o	τ_a
7-Apr-2012												
652	0.184	0.003	0.051	0.143	0.009	0.048	0.097	0.024	0.042	0.046	0.014	0.031
655	0.184	0.003	0.046	0.143	0.009	0.044	0.097	0.024	0.038	0.046	0.014	0.028
658	0.184	0.003	0.056	0.143	0.009	0.053	0.097	0.024	0.046	0.046	0.014	0.035
701	0.184	0.003	0.054	0.143	0.009	0.051	0.097	0.024	0.044	0.046	0.014	0.033
707	0.184	0.003	0.055	0.143	0.009	0.051	0.097	0.024	0.044	0.046	0.014	0.034
710	0.184	0.003	0.052	0.143	0.009	0.049	0.097	0.024	0.042	0.046	0.014	0.032
713	0.184	0.003	0.051	0.143	0.009	0.048	0.097	0.024	0.041	0.046	0.014	0.031
716	0.184	0.003	0.046	0.143	0.009	0.043	0.097	0.024	0.037	0.046	0.014	0.028
719	0.184	0.003	0.045	0.143	0.009	0.042	0.097	0.024	0.037	0.046	0.014	0.027
725	0.184	0.003	0.048	0.143	0.009	0.045	0.097	0.024	0.038	0.046	0.014	0.029
728	0.184	0.003	0.052	0.143	0.009	0.048	0.097	0.024	0.042	0.046	0.014	0.032
731	0.184	0.003	0.043	0.143	0.009	0.040	0.097	0.024	0.035	0.046	0.014	0.026
734	0.184	0.003	0.046	0.143	0.009	0.043	0.097	0.024	0.037	0.046	0.014	0.028
737	0.184	0.003	0.045	0.143	0.009	0.042	0.097	0.024	0.036	0.046	0.014	0.027
740	0.184	0.003	0.048	0.143	0.009	0.045	0.097	0.024	0.039	0.046	0.014	0.029
746	0.184	0.003	0.047	0.143	0.009	0.044	0.097	0.024	0.038	0.046	0.014	0.029
749	0.184	0.003	0.059	0.143	0.009	0.056	0.097	0.024	0.048	0.046	0.014	0.037
755	0.184	0.003	0.075	0.143	0.009	0.075	0.097	0.024	0.066	0.046	0.014	0.051
...	...	...	...	...	...	...	...	...	...	...	...	...

τ_R: Rayleigh optical depth, τ_o: ozone optical depth, τ_a: aerosol optical depth

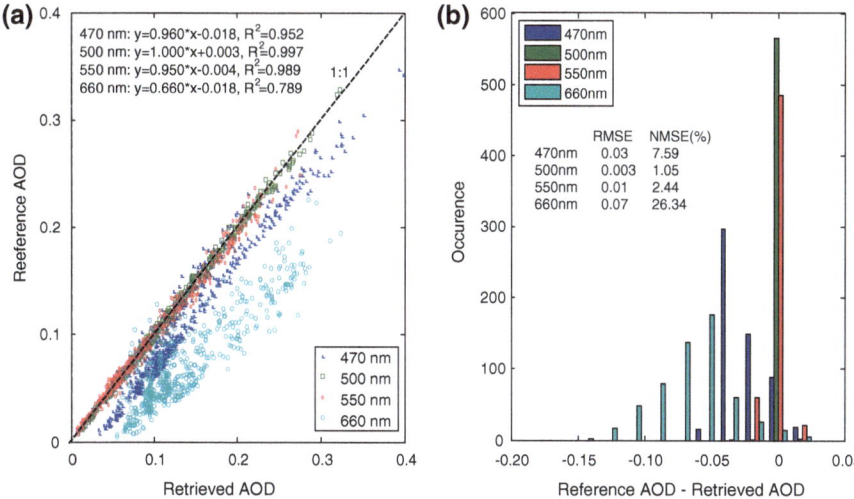

Fig. 4.6 Comparison between retrieved and reference AOD in irradiance-match validation using i-SMARTS in (**a**) scatter plot and (**b**) histogram plot. (RMSE: root mean square error, NMSE: normalized mean square error)

with $R^2 > 0.95$ and RMSE 0.028. Besides, overall inspection in the error analysis in Fig. 4.7b also suggests that overestimation is likely to happen for both 470 and 660 nm. However, for wavelength 500 and 550 nm, the statistical analysis verifies that the retrieved AOD closely matches the reference values with high correlation $R^2 > 0.98$ and low RMSE < 0.01.

Table 4.7 shows the conclusive results between the measured and calculated irradiance in RCE validation as well as the retrieved and reference AOD in i-SMARTS validation. Since the retrieved AOD is validated against the reference AOD that is primarily dependent on the accuracy of measured irradiance, therefore the total uncertainty in AOD retrieval at distinct wavelengths is established from the error attributed to the measured irradiance plus the error resulted from the AOD comparison. AOD retrieval in mid-visible range (500 & 550 nm) has a total uncertainty ~ 5 % but errors in the edge of the range (470–660 nm) are relatively larger at ~ 9 % and ~ 27 %. This further indicates the uncertainty of AOD measurement from the proposed algorithm is wavelength dependent by the most at near-infrared bands.

Figure 4.7 and Table 4.8 present the daily analysis of zero airmass (AM0) extrapolation Langley plot at multiple wavelengths within the observation period. The daily extrapolated value varies in the range of 9.64–10.38 at 470 nm, 9.64–10.27 (500 nm), 9.56–10.09 (550 nm), and 8.95–10.00 (660 nm). The maximum and minimum drifts are represented by red line and green line, respectively in Fig. 4.7 and bold values in Po column in Table 4.8. Consistently, they are all plotted from days that have few data of 3–9 points. While, other extrapolated lines that consisted of more data >10 tend to obtain a more consistent

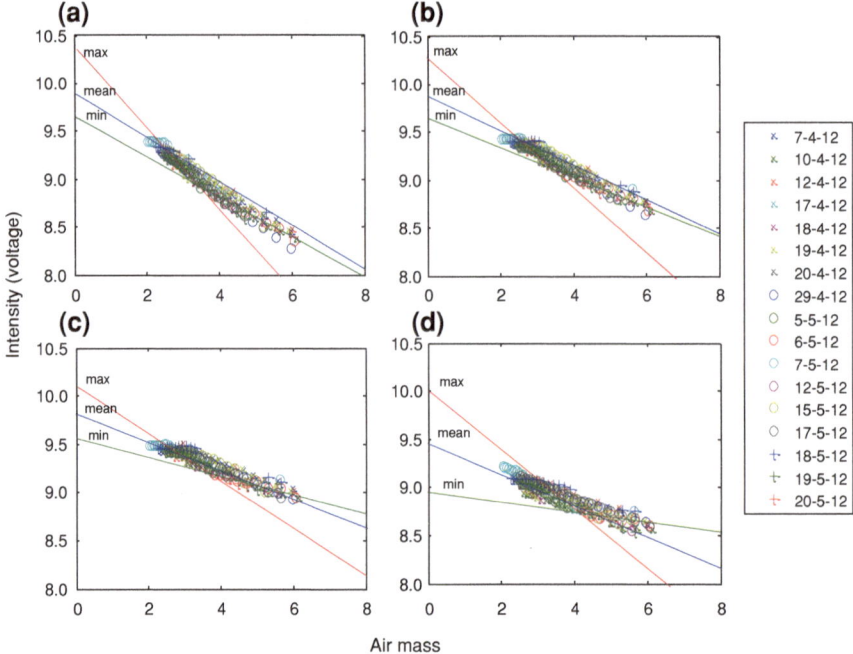

Fig. 4.7 Daily AM0 extrapolation Langley plot at **a** 470 nm, **b** 500 nm, **c** 550 nm, and (d) 660 nm. Legend box represents data in dd-mm-yy. The extrapolation line in *red*, *blue*, *green* represents maximum, mean and minimum calibration drifts within the observation period

Table 4.7 Result comparison between validation by radiative closure experiment (RCE) and irradiance-match using i-SMARTS over study area

Wavelength (nm)	Validation approach			
	RCE		i-SMARTS	
	RMSE	NMSE	RMSE	NMSE
470	0.11	1.95	0.03	7.59
500	0.16	3.11	0.003	1.05
550	0.16	3.25	0.01	2.44
660	0.08	0.74	0.07	26.34

RCE validates in DSI in W/m^2 /nm; i-SMARTS validates in AOD
RMSE: Root mean square error; NMSE: Normalized mean square error in %

extrapolated value close to the mean value. This indicates that the drifts are more likely to reduce when adequate useful data are available for the Langley plot. The calibration drifts are the least at wavelength 550 nm (std. dev. 0.12), followed by 500 nm (0.15), 470 nm (0.18) and 660 nm (0.24). This pattern could offer some explanations to the observed errors in AOD retrieval where it is apparent that wavelength 500 and 550 nm shows less calibration drifts compared to 470 and

Table 4.8 Daily analysis of Langley regression at multiple wavelengths. Symbol R^2, n, and P_o represents Langley correlation, number of data, and extrapolated value, respectively. The P_o value in bold represents the maximum and minimum calibration drifts

Date	470 nm				500 nm				550 nm				660 nm			
	Regression line	R^2	n	P_o	Regression line	R^2	n	P_o	Regression line	R^2	n	P_o	Regression line	R^2	n	P_o
07-04-12	$y = 0.24x + 9.89$	0.99	20	9.89	$y = 0.18x + 9.87$	0.99	20	9.87	$y = 0.12x + 9.80$	0.97	20	9.80	$y = 0.11x + 9.33$	0.97	18	9.33
10-04-12	$y = 0.25x + 9.86$	0.99	18	9.86	$y = 0.20x + 9.88$	0.99	18	9.88	$y = 0.15x + 9.82$	0.99	18	9.82	$y = 0.12x + 9.26$	0.97	18	9.26
12-04-12	$y = 0.22x + 9.85$	0.96	5	9.85	$y = 0.17x + 9.84$	0.95	5	9.84	$y = 0.14x + 9.84$	0.88	6	9.84	$y = 0.15x + 9.49$	0.89	14	9.49
17-04-12	$y = 0.29x + 10.09$	0.95	3	10.09	$y = 0.24x + 10.07$	0.93	7	10.07	$y = 0.16x + 9.87$	0.91	7	9.87	$y = 0.31x + 10.00$	0.97	4	**10.00**
18-04-12	$y = 0.26x + 9.88$	0.97	17	9.88	$y = 0.20x + 9.86$	0.97	16	9.86	$y = 0.14x + 9.78$	0.96	16	9.78	$y = 0.13x + 9.31$	0.96	10	9.31
19-04-12	$y = 0.22x + 9.77$	0.97	17	9.77	$y = 0.16x + 9.74$	0.93	18	9.74	$y = 0.11x + 9.70$	0.90	17	9.70	$y = 0.10x + 9.23$	0.93	14	9.23
20-04-12	$y = 0.42x + 10.38$	1.00	3	**10.38**	$y = 0.33x + 10.27$	0.83	5	**10.27**	$y = 0.24x + 10.09$	0.76	5	**10.09**	$y = 0.28x + 9.80$	0.91	4	9.80
29-04-12	$y = 0.29x + 9.98$	1.00	21	9.78	$y = 0.22x + 9.94$	0.99	21	9.94	$y = 0.15x + 9.86$	0.99	21	9.86	$y = 0.13x + 9.41$	1.00	21	9.41
05-05-12	$y = 0.34x + 10.15$	0.97	8	10.15	$y = 0.28x + 10.13$	0.95	10	10.13	$y = 0.21x + 10.02$	0.92	10	10.02	$y = 0.24x + 9.75$	0.91	10	9.75
06-05-12	$y = 0.21x + 9.64$	0.98	9	**9.64**	$y = 0.15x + 9.64$	0.97	9	**9.64**	$y = 0.10x + 9.56$	0.93	9	**9.56**	$y = 0.05x + 8.95$	0.64	6	**8.95**
07-05-12	$y = 0.27x + 10.00$	0.94	13	10.00	$y = 0.16x + 9.79$	0.94	14	9.79	$y = 0.11x + 9.74$	0.92	14	9.74	$y = 0.15x + 9.50$	0.89	14	9.50
12-05-12	n/a	n/a	1	n/a	n/a	n/a	1	n/a	n/a	n/a	1	n/a	$y = 0.17x + 9.49$	1.00	3	9.49

(continued)

Table 4.8 (continued)

Date	470 nm				500 nm				550 nm				660 nm			
	Regression line	R^2	n	P_o	Regression line	R^2	n	P_o	Regression line	R^2	n	P_o	Regression line	R^2	n	P_o
15-05-12	$y = 0.28x + 10.04$	0.99	20	10.04	$y = 0.21x + 9.96$	0.96	20	9.96	$y = 0.15x + 9.87$	0.92	20	9.87	$y = 0.16x + 9.50$	0.96	20	9.50
17-05-12	$y = 0.29x + 10.01$	1.00	17	10.01	$y = 0.21x + 9.92$	0.98	18	9.92	$y = 0.15x + 9.86$	0.97	20	9.86	$y = 0.15x + 9.47$	1.00	21	9.47
18-05-12	$y = 0.23x + 9.89$	0.98	10	9.89	$y = 0.16x + 9.83$	0.95	13	9.83	$y = 0.11x + 9.76$	0.88	15	9.76	$y = 0.11x + 9.38$	0.95	19	9.38
19-05-12	$y = 0.30x + 10.02$	0.99	18	10.02	$y = 0.23x + 9.96$	0.98	18	9.96	$y = 0.17x + 9.88$	0.98	19	9.88	$y = 0.12x + 9.46$	0.99	21	9.46
20-05-12	n/a	n/a	0	n/a	$y = 0.21x + 9.84$	1.00	2	9.84	$y = 0.15x + 9.77$	0.92	3	9.77	$y = 0.13x + 9.28$	0.94	4	9.28
Total	$y = 0.26x + 9.93$	0.96	200	9.93	$y = 0.19x + 9.87$	0.94	215	9.87	$y = 0.14x + 9.81$	0.90	221	9.81	$y = 0.14x + 9.43$	0.87	221	9.43
Max				10.38				10.27				10.09				10.00
Min				9.64				9.64				9.56				8.95
Mean				9.89				9.88				9.83				9.46
Std dev				0.18				0.15				0.12				0.24

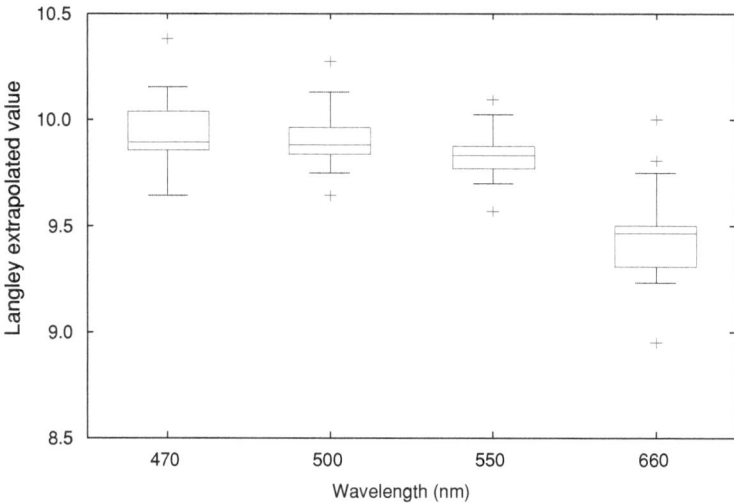

Fig. 4.8 Box plot of daily AM0 extrapolation Langley plot at 470 nm, 500 nm, 550 nm, and 660 nm. The *red* '+' represents outlier values

660 nm. It is well known that calibration drifts in Langley extrapolation has the integral effect on the retrieval of AOD particularly for high accuracy applications.

The definite reason behind the wavelength dependent calibration drifts is unknown but potentially justified by the limitation of the instrument for which could be partly due to the spectrally decrease in the photodiode sensitivity. Figure 4.8 visualizes the instrument's spectrally dependent limitation by plotting the Langley extrapolated values in boxplot for each wavelength. The deviation in Po varies in ascending order from 550, 500, 470 to 660 nm. For an ideal instrument response, the extraterrestrial constant should be invariant over time. Slight deviation in Po at 550 and 500 nm is indicative of highly stable instrumental response despite of varying measured voltage over time. On the other hand, considerable deviances at 470 and 660 nm indicate that these wavelengths do not possess the same attribute. On this ground, it is believed that instrument factors form potentially large error sources. The principal items needing attention are stray light and certain aspects of the optical design. This effect is even more severe when the stray light contaminations fictitiously contribute to the voltage reading of the instrument. The contamination results in a finite out-of-band rejection (OBR) which causes light of other more distant wavelengths than those specified by slit's function FWHM also contribute to the signal. It is common that stray light is undoubtedly a serious problem for some instruments particularly those without narrow FOV. The fact that the current instrument adopts merely a cosine corrector of 5° which is relatively less sensitive for direct solar irradiance measurement compared to other narrower FOV sunphotometers suggests that this effect is even more likely.

Thus, the overall effect leads to an erroneous estimation in the AOD retrieval particularly for wavelengths that fall outside the peak response of the instrument. Nonetheless, this error has minimum effects in the mid-visible range where the relative response of the instrument is the highest (see Fig. 4.1a). The total uncertainty of the AOD retrieval in this range is close to acceptable error ~ 5 %. Finally, at least for an overall inspection, AOD retrieval from the proposed algorithm is reliable in the mid-visible wavelengths on the grounds that the RMSE is comparable to the total uncertainty in AOD retrieval from a newly calibrated field instrument CIMEL/AERONET under cloud-free skies in visible range of ± 0.01–0.02 as stated by Holben et al. (1998).

References

Acker JG, Leptoukh G (2007) Online analysis enhances use of NASA Earth Science Data. Trans Am Geophys Union 88:14–17

Chang JHW, Dayou J, Sentian J (2014) Development of near-sea-level calibration algorithm for aerosol optical depth measurement using ground-based spectrometer. Aerosol Air Qual Res 14:386–395

Chang JHW, Dayou J, Sentian J (2013) Diurnal Evolution of Solar Radiation in UV, PAR and NIR Bands in High Air Masses. Nature, Environ Pollut Technol 12:1–6

Holben BN, Eck TF, Slutske I et al (1998) AERONET—a federated instrument network and data archive for aerosol characterization. Remote Sens Environ 66:1–16

Kaskaoutis DG, Kambezidis HD (2008) The role of aerosol models of the SMARTS code in predicting the spectral direct-beam irradiance in an urban area. Renew Energy 33:1532–1543. doi:10.1016/j.renene.2007.09.006

Knobelspiesse KD, Pietras C, Fargion GS et al (2004) Maritime aerosol optical thickness measured by handheld sun photometers. Remote Sens Environ 93:87–106

Seidel FC, Kokhanovsky AA, Schaepman ME (2012) Fast retrieval of aerosol optical depth and its sensitivity to surface albedo using remote sensing data. Atmos Res 116:22–32. doi:10.1016/j.atmosres.2011.03.006

Shettle EP, Fenn RW (1979) Models for the aerosols of the lower atmosphere and the effects of humidity variations on their optical properties. AIR FORCE Geophys, LAB HANSCOM AFB MA

Sumari SM, Darus FM, Kantasamy N et al (2009) Compositions of rainwater and aerosols at global atmospheric watch in Danum Valley, Sabah. Malaysia J Anal Sci 13:107–119

Trivitayanurak W, Palmer PI, Barkley MP et al (2012) The composition and variability of atmospheric aerosol over Southeast Asia during 2008. Atmos Chem Phys 12:1083–1100. doi:10.5194/acp-12-1083-2012

Utrillas MP, Martinez-Lozano JA, Cachorro VE et al (2000) Comparison of aerosol optical thickness retrieval from spectroradiometer measurements and from two radiative transfer models. Sol Energy 68:197–205

Chapter 5
Conclusion

Abstract Aerosol optical depth (AOD) is a measurement that represents the total attenuation of solar terrestrial radiation caused by aerosols. Measurement of AOD is often performed using ground-based spectrometers, because this approach has the highest accuracy, as well as high spectral and temporal resolutions. However, frequent calibration of ground-based spectrometers is often difficult. This is because conventional method usually not readily available for most users and also always complicated by possible temporal drifts in the atmosphere. The new Langley calibration algorithm has helped to provide solution to this issue to allow frequent calibration for ground-based spectrometers, even at near-sea-level sites that is comparable to conventional calibration approach performed at high altitude

Keywords Aerosol monitoring and measurements · Initial calibration · Broad-band pyranometer · Reliable calibration

5.1 Overview

Several methods can be used for aerosol monitoring and measurements. Among them, sunphotometers offer the most economical yet reliable method. To make this instrument more practical for these purposes, Langley calibration using Perez-Dumortier algorithm (or simply PDM) was developed so that the calibration procedure can be carried out at any altitude level. It is worth to mention again that the conventional Langley calibration is usually performed at high altitude observatory where homogenous, stable and clear sky is likely to happen. On the contrary, the PDM calibration algorithm makes the calibration feasible at any altitude level without travelling to high altitude sites as in conventional method always practiced. Most importantly, it requires no knowledge of instrument initial calibration or collocated calibrated instrument to transfer nominal calibration for cloud-screening. It is basically an ensemble combination of PDM clear-sky filtration model and statistical filter which serves as an objective algorithm to constrain the Langley extrapolation in getting the closest possible extraterrestrial constant over a wide range of wavelengths.

In original, Perez and Dumortier model were developed for broadband irradiances where in common practice broadband pryanometer alongside the sunphotometer is necessary. Again, this strongly limits the overall practicality of the method since very few sunphotometer stations are also equipped with broadband radiometers. To apply the models for this specific Langley-plot calibration, this additional broadband channel is unnecessary on the basis that the model calculation is merely based on the empirical ratio between the global and diffuse solar component. When broadband channel is unavailable, this ratio can actually be approximated by finding the ratio of the areas under the spectrum between these two components in spectral measurement using the trapezoid rule of integration. Thus, the application of the algorithm does not essentially depend on a collocated broadband radiometer.

Despite adequate removal of cloudy periods using PDM models, non-varying AOD values during the experiment is also another important condition for ideal Langley plot. Some works accept the time evolution of AOD during the day and compensate it on the Langley plot data (Campanelli et al. 2004). Other works do not compensate it but perform statistical analysis of the variations introduced in the Langley data, in order to remove the affected points (Harrison and Michalsky 1994). Although the approach is valid, still some parabolic evolution of the aerosol load during the day could affect the extrapolation of the calibration factor, and this effect would not be evident on the regression performed based on the assumption that these points are removed by reducing its coefficient of variation (CV) of the measurements throughout the experiment.

In many cases the temporal variability of the clouds particularly for calibration at near-sea-level makes it very difficult to select appropriate data suited for Langley plot in automated manner. Without an effective cloud-screening procedure, frequent calibration at near-sea-level is merely fictional. Either over-filtration or under-filtration may result to inappropriate data regression and thus contributes to fictitious extrapolation. With the aim to reduce the cloud contamination, the algorithm implements a strict cloud-screening procedure as an execution to select appropriate data from an extended period of measurements for the Langley-purpose calibration. Within the observation, AOD retrieval at 500 and 550 nm has a total uncertainty close to 5 % but at 470 and 660 nm the uncertainty is relatively higher than 9 and 27 %, respectively. The observed error is likely due to calibration drifts in the daily Langley extrapolated value where the drifts in mid-visible range wavelengths are relatively lesser (std. dev. <0.15) when compared to that of 470 nm (0.18) and 660 nm (0.24). Future improvements on this issue can be focused on integrating possible measures to control this wavelength dependent error.

The overall inspection suggests that the uncertainty of the PDM calibration method lies within the range of error ~5 % at mid-visible wavelengths. In comparative with most other related studies, the obtained result is compared with other normally calibrated spectrometers in Table 5.1. The uncertainty range found

Table 5.1 Results comparison with other normally calibrated spectrometer

Reference	Instrument	Calibration method	Uncertainty (%)
(Weihs et al. 1995)	MAINZ II Sunphotomter	Langley plot analysis at Zugspitze (2960 m)	3
(Reynolds et al. 2001)	Fast-Rotating Shadowband Spectral Radiometer (FRSR)	Langley plot analysis at MLO	2
(Augustine et al. 2003)	Multifilter-Rotating Shadowband Spectral Radiometer (MFRSR)	Langley plot analysis at the Table Mountain SURFAD stations by Long and Ackerman method	1–5
(Sano et al. 2003)	CE-318 and POM-100P Sunphotometer	Langley plot analysis at NASA/GSFC and MLO	<4
(Che et al. 2009)	CE-318 Sunphotometer	Langley plot analysis at Izana Observatory (AEMET, Spain) by secondary calibration method	<1.5
(Qiu 2010)	Pyrheliometer	In-situ measurement	±10
(Lee et al. 2010)	Multifilter-Rotating Shadowband Spectral Radiometer (MFRSR)	Langley plot analysis at Xianghe, Taihu, and Shouxian, China by Maximum Value Composite (MVC) method	<6
(Che et al. 2011)	CE-318 Sunphotometer	Langley plot analysis at Mt. Waliguan Observatory, China	0.5–1.0
Current method	ASEQ Spectrometer	Langley plot analysis at near-sea-level by PDM calibration algorithm	5

is comparable to most of the reported calibration in the literatures performed at high altitude and also well within the acceptable range of error. As an overall validation, the near-sea-level Langley calibration algorithm as is proven to be acceptable for reliable calibration of ground-based spectrometers for AOD measurement.

References

Augustine JA, Cornwall CR, Hodges GB et al (2003) An Automated Method of MFRSR Calibration for Aerosol Optical Depth Analysis with Application to an Asian Dust Outbreak over the United States. J Appl Meteorol 42:266–278

Campanelli M, Nakajima T, Olivieri B (2004) Determination of the solar calibration constant for a sun-sky radiometer: proposal of an in-situ procedure. Appl Opt 43:651–659

Che H, Wang Y, Sun J (2011) Aerosol optical properties at Mt. Waliguan Observatory. China. Atmos Environ 45:6004–6009. doi:10.1016/j.atmosenv.2011.07.050

Che H, Zhang X, Chen H et al (2009) Instrument calibration and aerosol optical depth validation of the China Aerosol Remote Sensing Network. J Geophys Res 114:D03206. doi:10.1029/2008JD011030

Harrison L, Michalsky J (1994) Objective algorithms for the retrieval of optical depths from ground-based measurements. Appl Opt 33:5126–5132

Lee KH, Li Z, Cribb MC et al (2010) Aerosol optical depth measurements in eastern China and a new calibration method. J Geophys Res 115:1–11. doi:10.1029/2009JD012812

Qiu JH (2010) A method for simultaneous broadband solar radiation calibration and aerosol optical depth retrieval. IGARSS. pp 1063–1066

Reynolds RM, Miller MA, Bartholomew MJ (2001) Design, Operation, and Calibration of a Shipboard Fast-Rotating Shadowband Spectral Radiometer. J Atmos Ocean Technol 18:200–214. doi:10.1175/1520-0426(2001)018<0200:DOACOA>2.0.CO;2

Sano I, Mukai S, Yamauo M et al (2003) Calibration and validation of retrieved aerosol properties based on AERONET and SKYNET. Adv Sp Res 32:2159–2164

Weihs P, Dirmhirn I, Czerwenka-Wenkstetten IM (1995) Calibration of sunphotometer for measurements of turbidity. Theor Appl Climatol 51:97–104

Index

A

Absolute calibration, 6
Absorbing, 2, 4, 6
Accumulation mode, 2, 3
AERONET, 7, 14–17, 27
Aerosol optical depth, 5, 9, 18, 28, 45, 50
Aietken mode, 2
Airborne radiometer, 6
Angstrom's exponent, 15, 29
Atmospheric aerosol, 3, 9

B

Beer-Lambert-Bouger's Law, 13, 14

C

Calibration factor, 20, 42, 44
Circumsolar radiation, 22, 23
Clearness-nebulosity index, 41
Cloud condensation nuclei, 4
Cloud-scattering effects, 34
Cloud-screening algorithm, 15
Coarse mode, 2, 15
Convective cloud, 5

D

Diffuse radiation, 16
Direct solar irradiance, 16
Dumortier model, 33, 34, 42, 44

E

Electrical field, 10
Electromagnetic field, 9
Extinction, 11–15, 18
Extraterrestrial irradiances, 6, 18
Extraterrestrial solar spectrum, 14

H

High altitude, 7, 18, 21, 31, 32

L

Langley calibration, 9, 18, 21–23, 25, 26, 33,
    40, 44
Langley extrapolation, 25, 32, 35
Langley method, 6, 7, 17, 18, 21–25, 27, 28,
    29
Langley plot, 17–21, 23, 24, 26, 27, 35, 36,
    41–43
Lidar, 6, 14
Liquid droplets, 1

M

Manmade aerosol, 1
Mauna Loa Observatory, 7
Mid-visible, 14
Mie theory, 9–12
Monodisperse, 2
Monte Carlo, 29

N

Natural aerosols, 1
Near-sea-level, 32, 40
Nuclei mode, 2, 3

O

Ozone, 19, 20, 44, 45, 50

P

Perez model, 33, 34
Polydisperse, 2, 15
Primary aerosols, 2

J. Dayou et al., *Ground-Based Aerosol Optical Depth Measurement*
Using Sunphotometers, SpringerBriefs in Applied Sciences and Technology,
DOI: 10.1007/978-981-287-101-5, © The Author(s) 2014

R
Radiative closure experiment, 50
Radiative forcing, 4
Radiative transfer model, 17
Rayleigh, 19, 20, 34, 44, 45, 50
Refractive index, 9, 13
Repetitive regression algorithm, 35, 36, 41, 42

S
Satellite data, 6, 14
Scattering, 2, 4, 6, 9–15, 18, 22, 23, 34, 44

Secondary aerosols, 2
Solar terrestrial radiation, 4
Solid particles, 1
Statistical filter, 25, 32
Stoke parameters, 11
Sunphotometry radiometer, 6, 14

T
Total irradiance, 16
Trace gases, 19, 20